Defense, Security and Strategy Series

MILITARY SATELLITES: ISSUES, GOALS AND CHALLENGES

DEFENSE, SECURITY AND STRATEGY SERIES

Military Satellites: Issues, Goals and Challenges
Abel Chirila (Editor)
2009. ISBN: 978-1-60741-238-0

Defense, Security and Strategy Series

MILITARY SATELLITES: ISSUES, GOALS AND CHALLENGES

ABEL CHIRILA
EDITOR

Nova Science Publishers, Inc.
New York

For permission to use material from this book please contact us:
Telephone 631-231-7269; Fax 631-231-8175
Web Site: http://www.novapublishers.com

LIBRARY OF CONGRESS CATALOGING-IN-PUBLICATION DATA
Available upon request

ISBN: 978-1-60741-238-0

Published by Nova Science Publishers, Inc. ✦ New York

CONTENTS

Preface		**vii**
Chapter 1	Improving Satellite Protection with Nanotechnology *Joseph Huntington*	**1**
Chapter 2	DOD is Making Progress in Adopting Best Practices for the Transformational Satellite Communications System and Space Radar but Still Faces Challenges *United States Government Accountability Office*	**19**
Chapter 3	Space Based Infrared System High Program and its Alternative *United States Government Accountability Office*	**53**
Chapter 4	Space Acquisitions: DOD's Goals for Resolving Space Based Infrared System Software Problems Are Ambitious *United States Government Accountability Office*	**77**
Chapter 5	Space Acquisitions: DOD Is Making Progress to Rapidly Deliver Low Cost Space Capabilities, but Challenges Remain *United States Government Accountability Office*	**97**
Index		**111**

PREFACE

This book explores the United States reliance on space operations for its security. Identifying vulnerabilities will allow us to apply our full range of capabilities to ensure space superiority and continued support to joint military operations across the spectrum of conflict. Space superiority is as much about protecting our space assets as it is about preparing to counter an enemy's space or anti-space assets. Space-based technologies and services permit people to communicate, companies to do business, civic groups to serve the public and scientists to conduct research. Thus, the United States is very reliant on satellites and will likely continue to be for many years to come. Today, small nations, groups, and individuals can acquire ground target data from commercial imagery sources, navigation and weather data from government-owned satellites and state of the art command and control capabilities through commercial communications satellites. These functions are also discussed in this book.

Chapter 1 - While the United States has enjoyed an historic advantage in space and the USAF has enjoyed the luxury of operating relatively unimpeded in this medium, the nation has become vulnerable to threats that could damage or disable its vital satellite constellations. This paper examines the threat to satellites posed by ground-based directed energy weapons and the state of satellite-related nanotechnology research and development to demonstrate the applicability for mitigating this type of threat. The paper argues that USAF leaders need to make continued research and development into nanotechnology for satellite applications an investment priority. This will require a long and expensive commitment, only some of which will pay off, but it is necessary if the USAF and the nation are to maintain space supremacy.

Chapter 2 - The Department of Defense (DOD) is working to achieve information superiority over adversaries and share information seamlessly among disparate weapons systems. Two programs envisioned as a part of this effort are Transformational Satellite Communications System (TSAT) and Space Radar. TSAT is designed to provide rapid worldwide secure communications with air and space systems—including Space Radar—through radio frequency and laser communications links. Space Radar is expected to provide global all-weather intelligence, surveillance, and reconnaissance, particularly in denied areas, for military, national intelligence, and civil users. Both TSAT and Space Radar will require major software development efforts and employ a significant number of experienced staff.

Chapter 3 - The U.S. relies on infrared satellites to provide early warning of enemy missile launches and protect the nation, its military forces, and allies. In 1996, the Department of Defense

(DOD) initiated the Space Based Infrared System (SBIRS) program to replace the nation's current missile detection system and provide expanded capabilities to support intelligence, surveillance, and reconnaissance missions. DOD expected to field SBIRS by 2004 at a cost of about $4.2 billion. However, over the past 11 years, SBIRS has proven to be technically challenging and substantially more costly. In an effort to stem cost increases and schedule delays, DOD has restructured the program multiple times, including revising program goals.[1] SBIRS is now estimated to cost over $10.4 billion, and the first satellite launch is expected in 2008. Because of continuing problems with SBIRS, DOD began a parallel alternative effort in 2006 known as the Alternative Infrared Satellite System (AIRSS), to compete with SBIRS and ensure that the nation's missile-warning and defense capabilities are sustained, or possibly provide a follow-on capability to SBIRS. You requested that we assess both SBIRS and AIRSS. As agreed with your office, with respect to SBIRS, we focused on the extent to which DOD is prepared to deliver the first two SBIRS satellites within revised cost, schedule, and performance goals. With respect to AIRSS, we examined the adequacy of DOD's decision to proceed with AIRSS as an alternative to SBIRS as well as whether DOD is attaining the knowledge it needs to position the program for success. To address these objectives, we reviewed schedule and funding information and performed our own analysis of cost and schedule projections using the contractor's 2006 cost performance report data. We also examined the resources committed and planned as well as users' needs for the competing effort. We presented our preliminary findings on SBIRS and AIRSS in briefings to your staffs in March 2007. This letter transmits the information provided in that briefing. We conducted our work between August 2006 and March 2007 in accordance with generally accepted government auditing standards. A copy of the briefing is enclosed.

Chapter 4 - To mitigate the SBIRS flight software problems, DOD has assessed various alternatives and developed a way to implement the software redesign and oversee its development. In April 2008, DOD approved the redesign effort, which addressed problems with the original design that affected the timing of stored programs, distribution of control between processors, and failure at the hardware interface level. Six review teams comprised of 70 personnel in all evaluated the designs to ensure the technical solutions, development approach, and readiness of test facilities were adequate. DOD and its contractor are now implementing the simplified architecture, developing new software, and testing elements critical to the integration and test of systems. DOD is also improving its program oversight and better managing the SBIRS development, by acting on the recommendations of an Independent Program Assessment; addressing weaknesses in management responsibility, accountability and organizational structure; and establishing a central execution team.

Chapter 5 - Since GAO last reported on DOD's ORS efforts in 2006, the department has taken several steps toward establishing a program management structure for ORS and executing research and development efforts. On the programmatic side, DOD provided Congress with a plan that lays out an organizational structure and defines the responsibilities of the newly created Joint ORS Office, and describes an approach for satisfying warfighters' needs. DOD has also begun staffing the office. On the research and development side, DOD has launched one of its TacSat satellites—small experimental satellites intended to quickly provide a capability that meets an identified need within available resources—and has begun developing several others. It has also made progress in developing interface standards for satellite buses—the platform that provides power, altitude, temperature control, and other support to the satellite in space—and continued its sponsorship of efforts aimed at acquiring low cost launch vehicles. Despite this progress, it is too early to determine the overall success of these efforts because most are still in their initial phases.

In: Military Satellites: Issues, Goals and Challenges
Editor: Abel Chirila, pp. 1-18
ISBN: 978-1-60741-238-0

Chapter 1

IMPROVING SATELLITE PROTECTION WITH NANOTECHNOLOGY*

Joseph Huntington

DISCLAIMER

The views expressed in this academic research paper are those of the author and do not reflect the official policy or position of the US government or the Department of Defense. In accordance with Air Force Instruction 51-303, it is not copyrighted, but is the property of the United States government and is not to be reproduced or published without the permission of the Air War College.

ABSTRACT

While the United States has enjoyed an historic advantage in space and the USAF has enjoyed the luxury of operating relatively unimpeded in this medium, the nation has become vulnerable to threats that could damage or disable its vital satellite constellations. This paper examines the threat to satellites posed by ground-based directed energy weapons and the state of satellite-related nanotechnology research and development to demonstrate the applicability for mitigating this type of threat. The paper argues that USAF leaders need to make continued research and development into nanotechnology for satellite applications an investment priority. This will require a long and expensive commitment, only some of which will pay off, but it is necessary if the USAF and the nation are to maintain space supremacy.

---

* This is an edited, excerpted and augmented edition of a Blue Horizons Paper, Center for Strategy and Technology, Air War College publication.

INTRODUCTION

> "The United States relies on space operations for its security, and this reliance may make us vulnerable in some areas. Identifying vulnerabilities will allow us to apply our full range of capabilities to ensure space superiority and continued support to joint military operations across the spectrum of conflict. Space superiority is as much about protecting our space assets as it is about preparing to counter an enemy's space or anti-space assets....We must protect our space assets."
>
> Chief of Staff--General John P. Jumper, August 2004[1]
> AFDD 2-2.1, 2 August 2004

The United States is very reliant on satellites and will likely continue to be for many years to come. Americans have come to depend on satellites and satellite services in order to conduct their daily activities. Space-based technology "enters homes, businesses, schools, hospitals and government offices through its applications for transportation, health, the environment, telecommunications, education, commerce, agriculture and energy. Space-based technologies and services permit people to communicate, companies to do business, civic groups to serve the public and scientists to conduct research."[2]

The United States Air Force (USAF) has become equally dependent, relying on satellites for the planning and conduct of almost every military mission, in peacetime and during war, not only for itself, but to support its sister services and coalition partners. Although space supported the preparation and conduct of operations during Operations Desert Shield and Desert Storm in 1990 and 1991, its significance has increased dramatically over the past several years. Combat operations in Afghanistan and Iraq demonstrate as never before, the operational importance of space to the joint warfighter. Space products and services are now significantly more capable, more abundant, and more integrated into all phases of combat operations.[3]

While the United States has enjoyed an historic advantage in space and has enjoyed the luxury of operating relatively unimpeded, the nation has become vulnerable to threats that could damage or disable its vital satellite constellations. This vulnerability of United States satellites became very real on January 11, 2007, when China successfully demonstrated its capability to destroy an on-orbit satellite. China launched a missile which intercepted and destroyed its FY- 1C weather satellite. On January 19, 2007, in an attempt assure the international community, Chinese Foreign Affairs spokesman Liu Jianchao said "there's no need to feel threatened about this....we are not going to get into any arms race in space."[4] The reality is there are many ways to deny, disrupt, or physically destroy satellite systems. These include attacking the ground stations via physical or computer network attack, employing denial and deception measures, jamming satellite communications equipment, physical attack of the satellites, or detonating a low-yield nuclear device in the atmosphere.[5]

Space operations and the ability to deny another country's freedom of access to space are no longer confined to global military powers. Today, small nations, groups, and individuals can acquire ground target data from commercial imagery sources; navigation and weather data from government-owned satellites; and state of the art command and control capabilities through commercial communications satellites.[6]

This paper will discuss how nanotechnology can be used to improve the design of United States satellites to mitigate those threats. The paper will look at potential improvements to generic satellite systems but its scope will not extend to investigating possible benefits to mission payloads. It will also analyze the ability of these nanotechnology-derived improvements to mitigate the threat posed to United States satellites by ground-based directed energy weapons.

The paper will begin by describing the threat to United States satellites posed by directed energy weapons. The paper will specifically examine both lasers and high-powered microwaves, describing what they are and how they may threaten satellites. It will then discuss what countries currently possess these capabilities and what capability adversaries might possess in the 2025 timeframe. From these discussions, the paper draws upon common satellite orbits to show the role orbits play in satellite vulnerability. It also looks at common satellite sub-systems and explores how ground-based directed energy weapons could these systems.

Several definitions for nanotechnology are explored. The paper explains how nanomaterials are different from materials at the micro- or macro-scale, and details the different properties nanomaterials possess.

The paper then explains how the work from nanotechnology research and development projects can be applied to satellite sub-systems, both today, and in the 2025 timeframe. Specifically, the paper discusses surface coatings that could be used to improve satellite thermal control and electrical conductivity, nanomaterials to improve the radiation hardness of commercial-grade microprocessors for use in satellites, and nanomaterials for hardening satellite structures.

The paper concludes with an assessment of the ability for these nanotechnology improvements to mitigate the threat posed by ground-based directed energy weapons. It makes recommendations for continued research and development in nanotechnology as one means for the USAF to maintain its dominance of space through and beyond 2025.

Directed Energy Weapon Threats to US Satellites

> "We simply cannot afford to defend against all possible threats. We must know accurately where the threat is coming from and concentrate our resources in that direction."[7] Edwin Land, founder of the Polaroid Corporation

United States space systems provide a host of critical capabilities to the nation and "the US military is dependent on the use of space capabilities in all types of warfare to maintain a combat advantage over our adversaries."[8] However, today's space infrastructure is largely unprotected and is vulnerable to attack by potential adversaries employing a variety of means which could damage or disrupt satellite operations, from simple jamming to physical destruction. To achieve space superiority and ensure uninterrupted use of United States space assets, the USAF conducts Defensive Counterspace Operations. Air Force Doctrine Document 2-2.1 defines Defensive Counterspace (DC S) Operations as operations to "preserve US/friendly ability to exploit space to its advantage via active and passive actions to protect friendly space-related capabilities from enemy attack or interference."[9] DCS operations include hardening space systems, either by physically hardening the structures to protect from attack or by using filters and shielding to protect the satellite from radiation effects.

Adversaries can use offensive counterspace techniques to degrade United States space capabilities. These can range from passive means, such as denial and deception, to more active means, such as attacking ground or space components. Continued technological advances and proliferation of anti-satellite weapons will enable more adversaries to possess the means to attack or interfere with United States satellites.[10] Among these, directed energy weapons will provide adversaries means way to counter United States satellite operations, specifically using ground- based lasers and high-powered microwaves.[11]

Directed energy weapons offer the advantage of producing operational effects at the speed of light as well as the ability to engage multiple targets. Several nations, such as Russia and China, have either built or are developing the technology to construct ground-based directed energy weapons, either lasers or high-powered microwaves

Ground-based lasers could damage thermal control, structural and power system components and may affect electro-optical sensors on low earth orbiting satellites. Lasers generate and focus intense beams of light that can engage a target from a long distance. Low- power lasers are usually intended to spoof or jam satellite electro-optical sensors, resulting in temporary blindness of the satellite. High-power lasers cause damage or destruction by overheating parts of the satellite. Most susceptible are the satellite's structure, thermal control system, and solar panels.[12]

Long-range, ground-based high-power microwave systems are feasible and, in some cases, have application as potential anti-satellite weapons. The intense radiofrequency radiation from high-powered microwaves could disable or destroy sensitive electronic components.[13] High- power microwaves are likely to damage satellites using soft kill mechanisms, exploiting the satellite's inherent design vulnerabilities, rather than hard kill such as melting or blowing up the satellite. Soft kill damage can occur in one of two ways: in-band damage or out-of-band damage. Microwaves at the same frequency as the satellite's antennas enter the antennas and damage the internal circuitry by overloading them beyond their design limits with electromagnetic energy. Out-of-band damage occurs when microwaves enter through backdoors, or apertures not specifically designed as conduits for electromagnetic energy transmission. The resulting circuitry damage is from electromagnetically induced current resulting in thermal damage.[14] Thus, ground-based directed energy weapons present serious threats to United States satellites.

The question the United States military should ask is, what is the directed energy weapons threat today and how will it expand by 2025? In the near-term the principal threat is from two potential nations: Russia and China.

In a 1997 letter to President Clinton, Russian President Yeltsin acknowledged that at one time, Russia possessed an anti-satellite (ASAT) capability but that they renounced it when they realized the futility of a first-strike notion.[15] The renunciation aside, Russia still possesses ground-based laser systems capable of threatening United States satellites. The ground-based lasers at Sary Shagan in the south-central Soviet Union are capable of killing United States satellites at altitudes below 400 km and damaging satellites at altitudes up to 1,200 km.[16]

According to the DoD's 2005 report on "Military Power of the People's Republic of China," China is working on, and plans to field, ASAT systems. Chinese government officials have publicly indicated their intent to acquire radio-frequency weapons as a means of defeating technologically advanced military forces. China is also involved in advanced, state-of-the-art research and development of laser technologies and has fielded low energy laser weapons in its own forces. Non-weapon military lasers are already widespread in the PLA. Chinese writings suggest that radio-frequency and laser weapons could be used against satellites in orbit.[17]

China is conducting research to develop ground-based laser ASAT weapons and could eventually develop a laser weapon capable of damaging or destroying satellites. Whether China has tested such a capability is unclear.[18] However, in September 2006, the Pentagon released a statement saying "China could blind American satellites with a ground-based laser firing a beam of light to prevent spy photography as they pass over China."[19]

According to Jane's Intelligence Service, China is also believed to be developing high-powered microwave sources for RF weapons and is conducting research on electronics susceptibility to high-powered microwave pulses and atmospheric propagation.[20] This research activity should raise concern within the USAF space community.

Given the current state of technology, the constant rate of improvement and discovery, and globalization, ground-based directed energy weapons with the capability of damaging or destroying United States satellites will be more widely available by 2025. In addition to the threats posed by these countries' directed energy weapons, there is some evidence of increased danger due to technology sharing. Part of the former Soviet Union's significant investment in directed energy weapons may have found its way to China through personnel and business transactions. There is also evidence showing a significant level of Chinese-Russian co-operation on weapons development, making it possible that Russia may have transferred the knowledge to develop a nuclear-reactor powered, ground-based laser with ASAT capabilities to China.[21]

SATELLITE VULNERABILITIES TO DIRECTED ENERGY WEAPONS

> "There is a tendency in our planning to confuse the unfamiliar with the improbable. The contingency we have not considered looks strange; what looks strange is thought to be improbable; what is improbable need not be considered seriously."[22]
>
> Thomas Schelling, Political Economist

Satellites are vulnerable to attack. However, the signs of vulnerability are not always clear and therefore not always recognized. Hostile actions against space systems can reasonably be confused with natural phenomena and can be explained as computer hardware or software failure, even though it might actually result from a malicious act.[23] The question with regard to directed energy weapons then is, what about a satellite contributes to its vulnerability? Three things make a satellite vulnerable--its orbit, its design, and its electronics.

Satellites orbit the earth in a predictable fashion, due largely to their speed and altitude above the earth's surface. Depending on the mission, a satellite will generally be placed in one of three types of orbits--Low Earth Orbit (LEO), Medium Earth Orbit (MEO), or Geosynchronous Earth Orbit (GEO).

Satellites in LEO orbit at altitudes between 200-500 miles. This orbit is closest to the earth's surface, but satellites in LEO travel faster than in other orbits, relative to a point on the earth's surface, with speeds in excess of 17,000 miles per hour. A majority of earth sensing satellites (i.e., environmental sensing, intelligence, and imagery) are in this orbit because it allows them to be closer to the things they are sensing, making them more efficient. Compared to other orbits, satellites in LEO are more vulnerable due to altitude and less vulnerable due to speed. This makes them more difficult to find and track, but more susceptible to damage or destruction from directed energy weapons.

Satellites in MEO orbit at altitudes between 1,000-12,000 miles. This orbit is farther from the earth's surface, providing satellites with a larger field of view of the earth than a LEO satellite. In MEO at 12,000 miles above the earth, satellites travel at approximately 8,000 miles per hour relative to a point on the earth's surface. Currently, the United States uses this type of orbit for navigation systems (i.e., Global Positioning System).

Satellites in GEO orbit at 22,300 miles above the earth's surface. In this orbit, a satellite appears to remain positioned over a single point on the earth's surface. For this reason, GEO is widely used for communications and weather satellites. Relative to other orbits, satellites in GEO are less vulnerable due to altitude and more vulnerable due to speed. Their slow speed relative to a point on the earth's surface makes them easier to find and track, but their distance from the earth makes them more difficult for directed energy weapons to damage or destroy.

Given relatively predictable orbits, satellites are vulnerable to physical destruction by directed energy weapons. However, physically destroying the satellite itself, as the Chinese did in their January 2007 ASAT demonstration, isn't necessary. Damaging or destroying any of satellite's major systems would render it ineffective. Areas of vulnerability include solar panels, which provide a large surface area that is easily targetable; optical sensors, which are susceptible to "blinding;" and thermal control of the satellite and its components.

"The power system is the most critical system on any spacecraft because nearly every other subsystem requires power."[24] Solar panels, which usually consist of solar photovoltaic cells, convert solar energy into electrical energy. As a matter of design and physics, they present a large area relative to the rest of the satellite in order to collect as much solar energy as possible. As a result, their large surface area provides an equally large target for a directed energy weapon. It is possible that a laser could heat a satellite's solar cells so much that they lose their ability to conduct sunlight into electricity."[25] As a result, the satellite's power budget might be strained to the point that the batteries could not continue to support the satellite systems. It is also possible that the effect could be less dramatic, the result being a temporary power interruption during the attack but no permanent impact to the power system.[26] Such a capability, in effect, could allow an adversary to turn a satellite "off" without destroying it or causing permanent damage.

Sensing equipment (i.e., imagery) is susceptible to damage from directed energy weapons, usually by being overwhelmed by the incident energy. Lasers can damage or destroy optical instruments, such as charged coupled devices used to record imagery, which has the immediate effect of rendering the satellite incapable of accomplishing its mission.[27] Losing space-based imagery or weather data could have a serious effect on military campaigns.

Because satellites operate in the austere environment of space, thermal control is critical. There are several different sources of thermal energy acting on a satellite--solar radiation, earth- emitted infrared radiation, and heat generated by onboard equipment. In general, there are two types of thermal control systems, passive and active. A passive system relies on conductive and radiative heat paths and has no moving parts or electrical power input. An active system is used in addition to the passive system when passive system is not adequate, for example, on manned missions. Active systems rely on pumps, thermostats, and heaters, use moving parts, and require electrical power.[28] The thermal control systems are designed to keep sensitive components within very tight temperature tolerances, even as the space environment inflicts significant temperature fluctuations. A directed energy weapon used to illuminate a satellite could impede or destroy the thermal control system, making it difficult or impossible for the satellite to protect its critical components.

Satellite electronics are vital for satellite operation and mission accomplishment, but are susceptible to damage due to radiation effects of the space environment. This damage can present itself in one of two ways--soft errors or hard errors. Soft errors occur when a transient pulse or "latchup" in the device memory that causes a detectable error.[29] Latchup occurs when an ionizing trail generates a temporary electrical short circuit by creating a path between a current source and a current sink. The effects are usually reversible and temporary and can be cleared by removing and reapplying power to the affected circuit.[30] Hard errors may be physically destructive to the memory device and the effects tend to be permanent.[31] An example of a hard error would be a permanently destroyed microprocessor or memory device. It can be difficult, however, to determine whether the cause of these problems is due to something manmade or the result of continuous exposure to the space environment.

This vulnerability to space radiation makes satellite electronics equally vulnerable to manmade radiation sources, such as high-powered microwaves. The symptoms would be similar, making it difficult to immediately discern whether the satellite had been attacked. In effect, an adversary could damage the system while maintaining plausible deniability. As the USAF looks for ways to protect its satellites from ground-based directed energy weapons, it should know that possible solutions could also enhance day-to-day protection from the harsh environment of space.

WHAT IS NANOTECHNOLOGY?

"The pace of technological change is accelerating, and nanotechnology will be central to that change over the coming decades."[32]

K. Eric Drexler

While ground-based directed energy weapons pose a serious threat to United States satellites, there are emerging technologies whose methods could be used to improve existing technologies to enhance satellite protection. One problem with new technologies is that they are sometimes difficult to understand and often indistinguishable from magic.[33] In an attempt to lift the magic veil, this paper will discuss definitions of nanotechnology as well as information about the building blocks of the technology. The prefix "nano" corresponds to a basic multiplicand of 10^{-9}. In distance, a nanometer is hundreds or thousands of times smaller than a bacterium. Devices and systems at this size reach the limit of tens to hundreds of atoms.

Agreeing on a definition for nanotechnology isn't as easy as it may seem to be. Some subscribe to the view that nanotechnology involves anything that takes place at the sub-micron level. Sub-micron materials, such as hollow silica particles with diameters of 300 nm, have been used for more than 50 years and are now being relabeled as nanomaterials.[34]

Purists approach the definition differently. In 1987, K. Eric Drexler, writing in his book *Engines of Creation: The Coming Era of Nanotechnology*, coined the term nanotechnology as an analogy to microtechnology, which at the time was broadly applied to technology that manipulated matter at the micron scale. This seemed to work since nanotechnology would manipulate matter at the nanometer (nm) scale. To put the size of "nano" into perspective, a strand of human hair is approximately 75,000 nm wide. Ten hydrogen atoms placed in a row would measure approximately one nanometer in length. Generally, nanotechnology encompasses "materials that have one-dimensional properties or have properties that suddenly do not scale linearly with size reduction at the one-to-ten nanometer level."[35] A different definition characterizes nanotechnology as "a technology concerned with the production, study and utilization of lateral structures, layers, molecular units, inner boundary layers and surfaces with critical dimensions or production tolerances that extend from 100 nanometers down to atomic orders of magnitude."[36] Not to be outdone, the National Nanotechnology Initiative developed the following defining features of nanotechnology:[37]

1 Nanotechnology involves research and technology development at the 1 nm to 100 nm range.
2 Nanotechnology creates and uses structures that have novel properties because of their small size.
3 Nanotechnology builds on the ability to control or manipulate at the atomic level.

While there may not be total agreement about the definitions, there is wide agreement that nanoscale materials behave differently than they do at the macro- or microscales. An important reason for this difference is the significant change of material properties and physical phenomena. The first factor is that nanoscale dimensions approach characteristic, or quantum, waveform excitations in the material. The solid-state physics community has been exploring the properties of quantum wells in which one dimension, the growth dimension, is on the nanoscale.

The second factor is that nanostructures have a very high surface to volume ratio, which means no atom is very far from an atomic interface. Because of their chemical, electronic, and reactive properties, nanoparticles can be exploited to produce improved electronic, magnetic, optic, and biomaterials.

Definitions aside, there are two different schools or approaches for "creating" nanotechnology material. These are the "top down" and the "bottom up" approaches. The top down approach involves reducing the structure sizes of microscopic objects to the nanometer scale using machining or etching techniques, the motivation for which is determined by microelectronics where sub-micrometer processes are being developed to move toward nanoelectronics for the next generation of electrical components. The bottom-up approach uses the controlled assembly of atomic and molecular elements to create larger systems. The bottom- up method has led to the development of several self-assembly techniques which can be used to form nanostructured layers or clusters.[38]

Carbon-base materials are ideal as molecular building blocks for nanoscale systems because carbon exists in a variety of forms and provides the basic shapes needed to build complex molecular-scale architectures (i.e., planar sheet, rolled-up tubular, helical spring, rectangular hollow box, conical, etc.). One of the structures most commonly identified with nanotechnology is the carbon nanotube. First discovered in 1991, carbon nanotubes have spawned science and engineering research devoted entirely to carbon nanostructures and their applications due in large part to the combination of their structural perfection, small size, high stiffness, high strength, and excellent electronic properties. Carbon nanotubes are tubular structures of carbon, in which each carbon atom is positioned in a lattice that wraps into a hollow pipe with a diameter from a few to tens of nanometers and can be either single-wall or multi-wall. Single-wall carbon nanotubes are best described as a rolled-up tubular graphene sheet composed of benzene-type hexagonal rings of carbon atoms. Multi-wall carbon nanotubes are multiple concentric single-wall carbon nanotubes. These two structures offer several interesting properties.

First, single-wall carbon nanotubes can be either metallic or semi-conducting, depending on the chiral vector, or amount of twist, of the lattice structure.[39] A carbon nanotube is metallic if electrons can freely move to the conduction band. A semi-conducting carbon nanotube requires additional energy before electrons can move to the conduction band. This has made them a candidate material for potential applications such as nanoscale devices and quantum wires. Researchers have demonstrated working carbon nanotube transistors which are a hundred times smaller than the 130-nanometer transistor gates currently found in computer chips and collections of nanotube transistors working together as simple logic gates.[40]

Single- and multi-wall carbon nanotubes display good elastomechanical properties because the carbon atom arrangement allows for large out of plane distortions while maintaining its strength when subjected to in-plane forces. This means nanotubes will bend under an extreme amount of force, but will return to their original shape when released. The Young's modulus for carbon nanotubes, a measurement of how much force it takes to bend a material, is about 5 times higher than for steel.[41] These characteristics point toward possibly using nanotubes in extremely lightweight, highly elastic and strong composite materials. However, while they display extraordinary resilience and flexibility at the nanoscale, it's not clear that these properties translate to the macroscopic scale.[42]

Carbon nanotubes conduct electricity better than metals because electrons traveling through carbon nanotubes follow quantum mechanical rules. Electrons exhibit ballistic transport, essentially behaving like a wave traveling through a smooth channel with no atoms to interfere with their motion. Ballistic electron transport, supported by many studies, is considered one of the reasons that nanotubes exhibit high current density when compared with other materials at similar nanoscale. This has resulted in considerable enthusiasm over the possibility of using carbon instead of silicon in the field of nanoelectronics. Multi-wall carbon nanotubes can pass a very high current density, from 10^6 to 10^8 amperes/cm^2, without suffering adverse effects. However, long-term stability while operating at these current densities remains a question.[43] As conventional CMOS electronics will soon reach economical and physical limits, nanoelectronic technologies may provide the basis for continued scaling of electronics into the next decade, following Moore's Law, and may provide the potential for hybrid architectures combined with traditional electronics.[44]

Carbon nanotubes also have high thermal conductivity, which means they conduct heat well. Unlike metals that depend on moving electrons to conduct heat, carbon nanotubes use the vibration of the covalent bonds that hold the carbon atoms together to conduct heat. Movement of the atoms themselves transmits heat throughout the material. Research has shown the thermal conductivity of single-wall carbon nanotubes to be only second to diamond, which has the highest thermal conductivity of any material.[45]

Public research and development programs for nanotechnology are a worldwide effort and quickly building momentum. A 2005 article in the *Journal of Nanoparticle Research* indicated that at least 60 countries have initiated nanotechnology initiatives and that "worldwide investment in nanotechnology research and development reported by government organizations has increased approximately nine-fold in the last eight years, from $432 million in 1997 to about $4.1 billion in 2005."[46] In the USAF, approximately 70 percent of the Air Force Office of Scientific Research's basic research funds are put toward this work, with the remaining 30 percent going toward Air Force Research Laboratory (AFRL) research programs.[47] The corporate world is also investing in nanotechnology. According to a report by Innovest Strategic Value Advisors, an investment group that closely monitors nanotechnology, advanced materials and nanotechnology investments increased from $68.2 million in the first quarter of 2004 to $83.5 million in the first quarter of 2005, a 22.4 percent increase."[48]

While much of the research has focused on the technology's physical properties and potential applications, some experiments have produced results that have raised concerns, especially since many products using nanomaterials are already on the market and more are on the way. Potential hazards to human health from the uncontrolled release and inhaling of nanoparticles or nanomaterials has thus far been based on analogies and the results of studies on the effects of ultrafine particles. In a DuPont study, fifteen percent of a rat population died due to lesions caused by nanotube clumps in their lungs. In a separate study conducted by Southern Methodist University, researchers found that buckyballs can disrupt fish brain cell membranes.[49] Though by no means conclusive, these studies and others highlight the fact that there's still much to learn. To that end, the Environmental Protection Agency has plans to lead a continued, comprehensive research effort to determine the risks manufactured nanomaterials pose to humans and the environment due to their composition, reactivity, and unique size.[50]

Applying Nanotechnology to Satellites

Given that the United States is highly dependent on satellites for daily life and military operations and that directed energy weapons pose serious threats to those satellites, the United States must harness emerging technologies to mitigate the threat. Nanotechnology may provide solutions that will enable the United States to mitigate the threat from directed energy weapons.

Hard, durable surfaces with coatings that can withstand extreme temperatures, abrasion, and wear are especially important for space vehicles. The satellite's surface would be one of the first parts to feel the impact of directed energy weapon's thermal or electromagnetic effects. Depending on its construction and design, the surface coating would either reflect, absorb, or transmit the incident energy or would perform some combination of the three. One-hundred percent reflection would be the ultimate protection because all the energy would be rejected; less than complete reflection would result in some absorption which would show as heat build-up, material degradation, or burn through.

Scientists can grow carbon nanotubes that exhibit reflective properties. These structures, called purified metal single-wall nanotubes, are grown as a mix of two isomers, metallic and semi-conducting.[51] The metallic nanotubes in the mixture have much better electrical conductivity and reflectance, but one of the biggest problems is the difficulty of separating the metallic nanotubes from the semi-conducting ones.[52] The resulting mixture isn't pure and must be further processed to remove the semi-conducting nanotubes and the other unwanted carbon particulates. When grown, the nanotubes are contaminated with a residual catalyst which must be removed in order to isolate the metal nanotubes. This is often accomplished using a strong acid which can also negatively alter defects on the nanotube surfaces and their reflectivity. This purification process induces major material and time costs to fabrication.[53]

While completely reflecting thermal and electrical energy would be preferred, dispersing it across the surface would also provide protection. The AFRL is managing a research program that uses carbon nanotube membranes, or Buckypaper, for electromagnetic shielding and to enhance lateral thermal conductivity. Buckypaper is a thin membrane, approximately 10-15 μm thick, of roped carbon nanotubes which are incorporated with composite structure.[54] The carbon nanotube ropes in Buckypaper are either randomly oriented or aligned in the same direction. The alignment determines the electrical and thermal properties. Randomly oriented Buckypaper has shown electrical conductivity in the range 45 0-670 Siemens per centimeter (S/cm) and thermal conductivity of 56 Watts per meter Kelvin (W/mK).[55] The electrical and thermal conductivities displayed by directionally aligned Buckypaper are 769-1,040 S/cm and 117 W/mK, respectively. A single layer of randomly oriented Buckypaper has an effective electromagnetic attenuation of 21dB/mil in the 4 to 12 GHz frequency range.[56] Buckypaper membranes are being investigated for aircraft lightning strike protection, but could have application to help satellites from electromagnetic events.[57]

Researchers have learned that vertically arranging carbon nanotubes in a sub-surface volume allows it to remove heat from the surface material. Using a process called Chemical Vapor Deposition, scientists can produce several square inches of vertically aligned carbon nanotube forests. In this form, carbon nanotube forests could be applied beneath a reflective coating on the satellite's surface for improved thermal dissipation. The commercial industry is pushing research teams to increase the size of the carbon nanotube forests they can grow because of the potential benefits as field emitters for the plasma screen television market. As such it is believed this process will be fully mature for commercial use by 2025.[58]

Carbon nanotube forests also offer a futuristic use that, though not tested, seems possible in theory. When energized with a small voltage at low pressure, carbon nanotube forests will emit electrons, which is the basis for their use as field emitters for plasma screen televisions. The emitted electrons ionize the atmosphere, generating a plasma shield around the structure. If the incident electromagnetic energy is short duration, the plasma should dampen most of the energy.[59]

In conjunction with the AFRL, the University of Dayton of Dayton Research Institute has developed a method to tailor the electrical conductivity of polymer materials used to build commercial and military aerospace components. This project transforms almost any polymer into a multifunctional material capable of carrying or dissipating significant electrical charge. Specifically designed carbon nanotubes with the current carrying capability of copper but at a much lower density, on the order of 50 to 150 nm in diameter, were carefully dispersed into a polymer matrix resulting in an electrically conductive polymer composite effective over the range 10^{-6} S/cm to 10^{2} S/cm. Researchers believe this technology is ready for commercialization and easily scalable to large batch production[60]

Slightly outside the pure definition of nano, the AFRL has been working with hollow silica particles, approximately 300 nm in diameter, as a pigment for satellite thermal control coatings. This technique uses the low refractive index of a void (i.e., an empty space) to promote energy scattering. The surrounding silica shell is transparent to ultraviolet light and is space-stable. The result is a broader spectrum of reflectance that extends into the ultraviolet frequency range and increased space durability because the particles don't absorb ultraviolet energy.[61]

Space presents an extremely harsh environment for satellites and the effects are especially severe for electronic components. Electronic components aboard a satellite in orbit around the earth encounter high-energy particle radiation many orders of magnitude higher than do similar components on Earth. The future of space nanoelectronics lies with a combination of commercial and radiation-hardened Complimentary Metal-Oxide Semiconductor (CMOS) circuits.[62] "Researchers are adapting commercial designs for space use with minor loss of performance, and modifying commercial processes to improve the radiation tolerance of components."[63]

The Defense Advanced Research Projects Agency (DARPA) and the National Reconnaissance Office (NRO) are working on similar efforts to improve the radiation hardening of CMOS circuits for use in space. Under a two-phase program called Radhard by Design, DARPA; partnering with the DoD Trusted Foundary Program which enables access to leading- edge semiconductor technologies, design libraries and design circuit cores; plans to develop and demonstrate techniques for fabricating strategically radiation hardened integrated circuits. Phase One seeks to demonstrate the efficacy from a technology standpoint and has the goal of getting circuits to withstand radiation total doses of 2 Mrad and dosage rates of 10^{10} rad(Si)/sec; with less than 10^{-10} errors/bitday. Phase Two will refine the data and techniques learned during Phase One to design and construct test structures and integrated circuit devices.[64]

In a similar effort, the NRO's Advanced Science and Technology Space Program Office is conducting a research and development effort that uses carbon nanotubes to enable a one gigahertz microprocessor capable of surviving the highest anticipated radiation levels. Circuit fabrication is compatible with existing CMOS processes and takes advantage of the electrical properties and strength of carbon nanotubes to create an electromechanical switch, a technique called Complementary Carbon Nanotube Logic (CCNL).[65]

CCNL enables switching by applying an electric field between the read/write electrode and the carbon nanotube fabric causes deforms the fabric until it contacts the read/write electrode, which turns the device "on." At this point the electric field can be turned off--van der Waals forces hold the carbon nanotube fabric in place. Reapplying an electric field is required to release the contact with the electrode. CCNL provides three key results. First, the switched bits are nonvolatile. Second, because van der Waals forces hold the carbon nanotube fabric in contact with the electrode, there is no leakage current and no standby current is needed. Lastly, CCNL makes it possible for one electromechanical switch to replace six to twelve memory transistors.[66] Based on the current program timeline, radiation-hardened microprocessors that use CCNL will be ready for space flight qualification in the Fiscal Year 2011 timeframe, as long as the project continues on its current path, with continued funding and no major setbacks.[67]

In addition radiation hardening, nanomaterials are also being investigated for their potential to protect satellites from electromagnetic interference. Through research, Northrop Grumman has demonstrated the feasibility of using nickel nanostrands as an electromagnetic shield for satellites.[68] Nickel nanostrands are made from strands of sub-micron diameter nickel particles that are linked in chains from microns to millimeters in length. They are very similar to multi-wall carbon nanotubes, but have the electromagnetic, chemical, and metallurgical properties of nickel.[69] Electromagnetic shielding that uses nickel nanostrands performs almost as well as current carbon fiber and aluminum shielding and nickel is proving to be an easy material to work with. There are, however, some drawbacks. Because it's a new material for this type of application, it could potentially require new processes. Also, nickel is a heavy metal which is leading to concerns about possible toxicity from particles released during the milling process.[70]

CONCLUSION

> "New anti-access threats and threats to our military, civil, and commercial space systems are creating new challenges to overcome. In order to continue providing our forces, and those of our allies, the level of support on which they depend, we must modernize our space forces and pursue truly transformational capabilities."[71]
>
> General Lance Lord

The United States has long been dominant in space. Controlling the "high ground" has provided information superiority and significant force enhancement capabilities to the military.[72] This dominance has overshadowed the vulnerability of United States satellites. By studying United States capabilities, other countries have become increasingly aware of the tremendous military advantages that space-based assets provide the United States and are beginning to challenge the long-held peaceful use of space.[73] The January 2007 Chinese ASAT demonstration is the latest example of how easily US space dominance can be threatened. Other examples were highlighted in a National Air and Space Intelligence Center report which stated that China and Russia have either built or are developing the technology to build ground-based directed energy weapons.[74] By 2025, continued technological discovery and improvement, globalization, and technology sharing will contribute to increasing the threat, making these weapons equally available to state and non-state actors.

Ground-based directed energy weapons provide adversaries with the means to attack United States satellites. Though an adversary may have many reasons or objectives for taking such action, two objectives are plausible. First, an adversary could use ground-based directed energy weapons to destroy United States satellites in order to degrade or deny United States capabilities, military, economic, or informational. Successfully destroying a satellite in this manner might accomplish an adversary's objectives, but it would also leave evidence that would allow the United States to trace the attack to the adversary. This would almost certainly result in some sort of retaliatory action. Second, an adversary could choose to degrade a United States satellite or satellite capability over time. This might be accomplished by employing ground-based directed energy weapons over some period of time at lower output power levels than necessary for complete destruction. Because many of the effects experienced by the satellite would be similar to the effects encountered as a result of operating in space, it would be difficult for the United States to pinpoint the cause. Not only would the adversary achieve the desired effect, he would also be able to maintain plausible deniability.

Does nanotechnology help protect satellites against these attacks? According to some researchers and available literature, nanotechnology and nanomaterials can technically be used to protect satellites against the effects of directed energy weapons. However, this has yet to be sufficiently demonstrated. The research and testing of this specific nanotechnology application is in its early stages. Much more work needs to be done to understand how the unique properties of nanomaterials can be harnessed to protect satellites against directed energy weapons threats. This will require the USAF to make this research thrust a priority and maintain it as such, which translates into a long-term financial commitment. Recent comments by Congressional members in the aftermath of the Chinese ASAT demonstration may help raise the priority of this research.[75]

Though nanotechnology isn't yet mature enough to be considered a solution for protecting satellites from ground-based directed energy weapons, the USAF shouldn't ignore it. On the contrary, the USAF should continue investing in nanotechnology research and development to understand and harness its capabilities for protecting critical satellite systems because nanotechnology will have a significant impact on future satellite design.

Though most nanotechnology work is still in basic research (i.e., 6.1), it clearly is a high-risk/high-payoff and transformational space capability critical to continued space supremacy. By 2025 it will have a significant impact on United States satellites, touching on the structure and functions of all satellites in the form of radiation-hardened microprocessers, enhanced surface coatings, and reduced satellite size and weight.

In the near-tem, radiation-hardened electronics and surface coatings are likely to provide the greatest benefits toward protecting satellite systems from directed energy weapons. Structural enhancements derived from macro-scale nano-structures are likely at least 10-15 years away.

Nanotechnology research will likely be more relevant for the United States government than for the commercial sector. As with all research, some of the efforts will pay off and some won't, but it's too early to know which areas will have success. Therefore, USAF must make continued research and development into nanotechnology for satellite applications an investment priority and must pay equal attention to all of these areas. The USAF will need to guide the research effort to ensure work is done to meet its stated needs and requirements. Undoubtedly, breakthroughs will happen, but not necessarily in areas where the USAF is invested or currently knows much about. This will require the USAF to stay current with nanotechnology research and development efforts around the world, in the academic, commercial, and governmental arenas, and be prepared to adapt the results to pursue applications critical for the nation's space systems. This means a long and expensive commitment, but one that's necessary if the USAF and the nation are to maintain space supremacy. Failure to do so will leave the nation in a position from which it will be difficult, if not impossible, to recover.

REFERENCES

[1] Jumper, General John P., *Air Force Doctrine Document 2*, Washington DC, August 2, 2004, paragraph 2.1.
[2] *Report of the Commission to Assess United States National Security Space Management and Organization*, Pursuant to Public Law 106-65, January 11, 2001 (Washington, DC), 11-12.
[3] Lieutenant General Larry J. Dodgen, "Space--Enabling the Potential of Our Joint Warfighter," *Quest for Space Magazine*, 2005, http://www.smdc.army.mil/pubaff/05Press/Space.html (accessed January 24, 2007).
[4] Richard Fisher, Jr., "China's Direct Ascent ASAT," International Assessment and Strategy Center, January 20, 2007, http://www.strategycenter.net/research/pubID.1 42/pub_detail.asp# (accessed January 22, 2007).
[5] *Report of the Commission to Assess United States National Security Space Management and Organization*, pp.19-21.
[6] *Report of the Commission to Assess United States National Security Space Management and Organization*, p.19.
[7] Conversation between Edwin Land and U.S. government analyst Albert Whelon. Reported in Whelon, Albert D. "Corona: The First Reconnaissance Satellites," *Physics Today*, February 1997, pp. 24-30.
[8] Air Force Doctrine Document (AFDD) 2-2.1, *Counterspace Operations*, 2 August 2004, p. 1.
[9] Ibid., p. 3.
[10] NASIC-1441-3894-05, *Challenges to US Space Superiority*, (Wright-Patterson AFB, OH: National Air and Space Intelligence Center, March 2005), pp. 21-25.
[11] While potential adversaries might also develop and employ air- or space-based systems, their effects would be mostly similar. Therefore, this paper will, for simplicity, concentrate on ground-based systems.
[12] Federation of American Scientists, "Threats to United States Space Capabilities," http://www.fas.org/spp/eprint/article05.html#17, (accessed December 1, 2006).
[13] Ibid.
[14] Phillip E.Nielsen, *Effects of Directed Energy Weapons*, (National Defense University, Washington, DC: 2003), http://www.ndu.edu/ctnsp/Nielsen-EDEW.pdf, (accessed January 25, 2007), pp. 243-254.
[15] Bill Gertz, "Yeltsin Letter Reveals Anti-Satellite Weapons," *The Washington Times*, November 7, 1997.

[16] James G. Lee, "Counterspace Operations for Information Domination," in *Beyond the Paths of Heaven: The Emergence of Space Power Thought*, ed. Bruce M. DeBlois (Maxwell AFB, AL: Air University Press, 1999), p. 283.
[17] Department of Defense, *Annual Report to Congress: Military Power of the People's Republic of China 2006*, (Washington, DC), pp. 34-35.
[18] Department of Defense, *Annual Report to Congress: The Military Power of the People's Republic of China, 2005*, (Washington, DC), p. 36.
[19] Francis Harris, "Beijing Secretly Fires Lasers to Disable US Satellites," *Telegraph*, September 26, 2006, http://www.telegraph.co.uk/news/main.jhtml?xml=/ news/ 2006/09/26/wchina226.xml, (accessed October 14, 2006).
[20] Jane's Intelligence Service, "China's Aerospace and Defence Industry, December 1, 2000", http://www8.janes.com/Search/documentView.do?docId=/content1/janesdata/srep/srep085/s0850010.htm@current&pageSelected=allJanes&keyword=directed%20energy%20weapons%20sensors&backPath=http://search.janes.com/Search&Prod_Name=SREP085& (accessed November 16, 2006).
[21] Ibid.
[22] Thomas Schelling was quoted by Roberta Wohlstetter in her1962 book. See: Wohlstetter, Roberta, *Pearl Harbor: Warning and Decision*, Stanford University Press, Stanford, CA; 1962, p. viii
[23] *Report of the Commission to Assess United States National Security Space Management and Organization*, p. 23.
[24] Texas Space Grant Consortium, "Spacecraft Design Archives," http://www.tsgc.utexas.edu/archive/subsystems/power.pdf (accessed September 27, 2006).
[25] Federation of American Scientists, *Ensuring America's Space Security: Report of the FAS Panel on Weapons in Space*, September 2004, p. 76.
[26] Ibid., pp. 78-79.
[27] Ibid., p. 76.
[28] Texas Space Grant Consortium, "Spacecraft Design Archives," http://www.tsgc.utexas.edu/archive/subsystems/thermal.pdf (accessed September 27, 2006).
[29] Kenneth LaBel et al., "Commercial Microelectronics Technologies for Applications in the Satellite Radiation Environment," http://radhome.gsfc.nasa.gov/radhome/papers/aspen.htm, (accessed December 8, 2006).
[30] National Research Council of the National Academies, *Implications of Emerging Micro- and Nanotechnologies* (Washington, DC: The National Academies Press, 2002), p. 53.
[31] LaBel et al., "Commercial Microelectronics Technologies for Applications in the Satellite Radiation Environment."
[32] Drexler, K. Eric, in Hall, Storrs J., *Nanofuture: What's Next for Nanotechnology* (Amherst, NY: Prometheus Books, 2005), Forward.
[33] Storrs J. Hall, *Nanofuture: What's Next for Nanotechnology*, p. 16.
[34] Dr Joel Johnson, scientist with the Air Force Research Laboratory (AFRL) Structural Materials Branch, Wright-Patterson AFB, OH, email correspondence with author, January 22, 2007.
[35] Ibid.

[36] Dr Wolfgang Luther, *International Strategy and Foresight Report on Nanoscience and Nanotechnology* (Düsseldorf, Germany, 19 March 2004), p. 3.

[37] Richard Booker and Earl Boysen, *Nanotechnology for Dummies* (Hoboken, NJ: Wiley, 2005), p. 10.

[38] Luther, *International Strategy and Foresight Report on Nanoscience and Nanotechnology*, pp. 4-5.

[39] Meyya Meyyappan and Deepak Srivastava, "Carbon Nanotubes" in *Handbook of Nanoscience, Engineering, and Technology*, ed. William A. Goddard, III, Donald W. Brenner, Sergey Edward Lyshevski, and Gerald J. Iafrate (Boca Raton, FL: CRC Press, 2003), p. 3.

[40] National Research Council of the National Academies, *Implications of Emerging Micro- and Nanotechnologies*, p. 48.

[41] Booker, *Nanotechnology for Dummies*, p. 77.

[42] Erik T. Thostenson and Tsu-Wei Chou, *Carbon Nanotube-Based Composites for Future Air Force and Aerospace Systems*, AFRL-SR-AR-TR-06-0165 (Arlington, VA: Air Force Office of Scientific Research, 2006), p. 6.

[43] B.Q. Wei, R. Vajtai, and P.M. Ajayan, "Reliability and Current Carrying Capacity of Carbon Nanotubes," *Applied Physics Letters* 79, no. 8 (20 August 2001): p. 1172.

[44] Luther, *International Strategy and Foresight Report on Nanoscience and Nanotechnology*, 8.

[45] Meyyappan, "Carbon Nanotubes," pp. 18-23.

[46] M.C. Roco, "International Perspective on Government Nanotechnology Funding in 2005," *Journal of Nanoparticle Research* 7, no. 6 (December 2005): p. 3.

[47] National Research Council of the National Academies, *Implications of Emerging Micro- and Nanotechnologies*, 184-185.

[48] Innovest Strategic Value Advisors, "Nanotech Benefits and Potential Risks: Innovest Launches Nanotech Index for the Value Investor," September 12, 2005, http://www.innovestgroup.com/pdfs/2005-09-15_Nanotechnology_PR.pdf, (accessed January 12, 2007), p. 2.

[49] Booker, *Nanotechnology for Dummies*, pp. 37-38.

[50] Environmental Protection Agency, *Nanotechnology: An EPA Perspective Factsheet*, (Washington, DC, March 2006), p.1.

[51] Dr Joel Johnson, scientist with the AFRL Structural Materials Branch, Wright-Patterson AFB, OH, email correspondence with author, January 26, 2007.

[52] Stephanie Reich, Christian Thomsen, and Janina Maultzsch, *Carbon Nanotubes: Basic Concepts and Physical Properties* (Weinheim, Germany: WILEY-VCH Verlag GmbH & Co, 2004), p. 86.

[53] Dr Andrey Voevodin, scientist with the AFRL Structural Materials Branch, Wright-Patterson AFB, OH, email correspondence with author, January 26, 2007.

[54] Dr Jennifer Fielding, program manager with the AFRL Structural Materials Branch, Wright- Patterson AFB, OH, email correspondence with author, January 19, 2007.

[55] Editor's note: The unit Siemens per centimeter (S/cm) is considered by many to be obsolete. However, this unit of measure was used by the author and is retained here. The more common unit, Siemens per meter, can be calculated by dividing the S/cm value by a coefficient of 100. Some texts refer to Watts per meter Celsius, instead of

Watts per meter Kelvin. As one Celsius degree is equal in magnitude to a Kelvin degree, the reader should note that these units are equivalent. .
[56] Florida Advanced Center for Composite Technologies, "SWNT Buckypaper Fact Sheet," January 2007, p. 1.
[57] Fielding, email correspondence, January 19, 2007.
[58] Dr Andrey Voevodin, email correspondence, January 26, 2007.
[59] Ibid.
[60] Air Force Research Laboratory, "Electrically Conductive Polymer Nanocomposite Materials," http://www.afrlhorizons.com/Briefs/Sept02/ML0206.html, (accessed January 22, 2007).
[61] Johnson, email correspondence, January 24-26, 2007.
[62] National Research Council of the National Academies, *Implications of Emerging Micro- and Nanotechnologies*, pp. 52-56.
[63] Air Force Research Laboratory, "Space Electronics the Radiation Hardened Way," http://www.afrlhorizons.com/Briefs/Sept01/VS0013.html, (accessed January 26, 2007).
[64] Defense Advanced Research Projects Agency (DARPA), "Radhard: Hardened by Design Overview," http://www.darpa.mil/mto/radhard/, (accessed January 24, 2007).
[65] Richard Ridgely, Program Manager for Nanotechnology Applications, National Reconnaissance Office, Advanced Science and Technology Space Program Office, Chantilly, VA, interview by the author, September 14, 2006.
[66] Richard Ridgley, "Leveraging Carbon Nanotubes to Develop a Fourth-Generation Radiation Hardened Microprocessor for Space," presented at the High Performance Embedded Computer Workshops, September 2006.
[67] Ridgley, interview.
[68] Dr Edward Silverman, Advanced Technology Manager, Northrop Grumman Corporation, Redondo Beach, CA, "Materials and Processes for Multifunctional Space Structures Program," June 29, 2006, pp. 11-14.
[69] Air Force Research Laboratory, "Nickel Nanostrands Expand Nanotechnology Design Engineering Capabilities," http://www.afrlhorizons.com/Briefs/Oct04/ML 0314.html, (accessed January 22, 2007).
[70] Dr Karla Strong, interview.
[71] Air Force Space Command Strategic Master Plan: *2006 and Beyond.*
[72] AFDD 2-2, 27 November 2001, p. 2
[73] David R. Tanks, *Future Challenges to U.S. Space Systems* (Washington, DC: The Institute for Foreign Policy Analysis, Inc., 1998), 1.
[74] NASIC-1441-3894-05, *Challenges to US Space Superiority*, p. 25.
[75] For more information on Congressional interest in ASAT protection, see Megan Scully, "House Republicans Call for Greater Military Effort in Space," Congress Daily, 1 February 2007, http://aimpoints.hq.af.mil/display.cfm?id=16399, (assessed February 1, 2007).

In: Military Satellites: Issues, Goals and Challenges
Editor: Abel Chirila, pp. 19-52
ISBN: 978-1-60741-238-0

Chapter 2

DOD IS MAKING PROGRESS IN ADOPTING BEST PRACTICES FOR THE TRANSFORMATIONAL SATELLITE COMMUNICATIONS SYSTEM AND SPACE RADAR BUT STILL FACES CHALLENGES*

United States Government Accountability Office

August 2, 2007
The Honorable Bill Nelson
Chairman

The Honorable Jeff Sessions
Ranking Member
Subcommittee on Strategic Forces Committee on Armed Services
United States Senate

The Honorable Ellen Tauscher Chairwoman

The Honorable Terry Everett Ranking Member
Subcommittee on Strategic Forces Committee on Armed Services
House of Representatives

The Honorable Silvestre Reyes
House of Representatives

The Department of Defense (DOD) is working to achieve information superiority over adversaries and share information seamlessly among disparate weapons systems. Two programs envisioned as a part of this effort are Transformational Satellite Communications System (TSAT) and Space Radar. TSAT is designed to provide rapid worldwide secure communications with air

* Excerpted from GAO Report GAO-07-1029R.

and space systems—including Space Radar—through radio frequency and laser communications links. Space Radar is expected to provide global all-weather intelligence, surveillance, and reconnaissance, particularly in denied areas, for military, national intelligence, and civil users. Both TSAT and Space Radar will require major software development efforts and employ a significant number of experienced staff.

TSAT and Space Radar development efforts are expected to be among the most costly space systems ever developed by DOD. In 2004, TSAT was estimated to have a total life cycle cost of about $16 billion, of which $2.0 billion will have been spent at the end of fiscal year 2007. Space Radar is estimated to have a total life cycle cost from $20 billion to $25 billion, and the program has spent about approximately $464.5 million. TSAT expects to begin product development in fiscal year 2008, and launch the first satellite in the first quarter of fiscal year 2016. Space Radar expects to begin product development in fiscal year 2009 and launch the first satellite in third quarter of fiscal year 2016. The systems are also expected to be among the most complex ever developed, largely because of the challenges associated with integrating critical technologies within the satellites and networking the satellites to other platforms.

You requested that we assess DOD's progress in adopting best practice as both of these programs proceed toward product development. We presented our findings on TSAT and Space Radar in briefings to your staffs in March 2007. This letter summarizes our findings, conclusions, and recommendations. Copies of the briefings are enclosed.

Results in Brief

DOD is making efforts to instill best practices on TSAT and Space Radar. These practices, as GAO has identified over the past decade, are to separate technology discovery from acquisition, follow an incremental path toward meeting user needs, match resources and requirements at program start, and use quantifiable data to make decisions to move to next phases. Collectively, these practices ensure a high level of knowledge is achieved at key junctures in development and that a program does not go forward unless a strong business case on which the program was originally justified continues to hold true.

However, sustaining these efforts could prove challenging. Specifically:

- Successful organizations we have studied ensure that technologies are mature, that is, proven to work as intended before program start. In the past DOD has chosen to extend technology invention into the acquisition process, and as a result, programs have experienced technical problems that require large amounts of time and money to fix. By contrast, best practice organizations mature technologies to the point of being tested in a relevant or operational environment before committing to an acquisition program.
- TSAT and Space Radar have made progress in the maturation of technologies, but challenges remain. In a June 2007 update, DOD determined that six of the seven critical technologies for TSAT are at a technology readiness level (TRL) 6 (meaning the technology has been tested in a relevant environment), and the program expects to have the remaining technology at a TRL 6 prior to the preliminary design phase. Space Radar expects to have almost all critical technologies mature to a TRL 6 by program start in June 2009. However, the program currently has five critical technologies assessed to be TRL 3

to TRL 4. This signifies that DOD will need to gain significant knowledge on these technologies to gain sufficient insight into costs and schedule to be well positioned for success by program start. In addition, the program office acknowledges that some of the seven technology risks it has rated as high, including risks related to spectrum, software, and integration with space radar users, will not be fully mitigated prior to program start.

- Successful organizations defer more ambitious technology efforts to corporate research departments (equivalent to the science and technology [S&T] organization in DOD) until they are ready to be added to future increments. Our best practice work has shown that a technology development environment is more forgiving and less costly than a delivery-oriented acquisition program environment. Events such as test failures, new discoveries, and time spent attaining knowledge are considered normal in this environment.
- Both programs have deferred more ambitious technology development efforts to the science and technology environment. TSAT, for example, deferred the inclusion of the wide-field of view multi- access communication technology to reduce risk on the program, and is currently contributing about $16.7 million for "off-line" maturation of this technology until opportunities arise for including it as part of future increments. In addition, it also eliminated multi-access laser communications[1] capabilities from consideration for future increments at this time due to the immaturity level of the technology. Space Radar has deferred lithium-ion batteries, more efficient solar cells, and onboard processing for its first increment, and like TSAT, is contributing toward their development by S&T organizations. At this time, Space Radar has not defined details of an increment beyond the first one.
- Successful organizations extensively research and define requirements before program start to ensure that they are achievable, given available resources, and that they do not define requirements after starting programs. In successful programs, negotiations and trade-offs occur before product development is started to ensure that a match exists between customer expectations and developer resources.
- Both programs have also strived to employ best practices to help identify and determine achievability of requirements. In 2006, the TSAT program was restructured into an incremental approach to control risk and increase confidence in the program schedule, putting agreements in place between development partners that organize capabilities into blocks based on technological maturity. For example, TSAT has reached agreements with groups representing the needs of users and warfighters that addresses which requirements will be included in the first and second blocks of the program. Space Radar has also developed an approach to obtain agreement and collaboration among users on program requirements. In an effort to facilitate communication and reach agreement over requirements between program partners within DOD and the Intelligence Community (IC), Space Radar has proactively introduced a variety of working groups that provide the program with a consolidated senior group of participants to validate, coordinate and integrate Space Radar requirements and concepts of operations throughout project development. Nevertheless, the Space Radar development effort has not yet had to fully define program requirements, including key performance parameters. Until all

[1] Multi-access laser communications technology is to provide simultaneous communications for a number of optical users at very high data rates.

requirements are defined, vetted, and validated, the program office could still face challenges in closing potential gaps between requirements and resources.

- Successful organizations ensure other resources—primarily funding, time, and people -- can also be matched to requirements before program start. Funding: Both programs face long-term challenges for funding. As DOD seeks to fund Space Radar and TSAT, it will be (1) undertaking other new, costly efforts, including the Global Positioning System III, the Space Based Surveillance System, and the Alternative Infrared Satellite System; (2) addressing cost overruns associated with legacy programs; and (3) facing increased pressures to ramp up investments in assets designed to protect space systems. In total, these efforts will increase DOD's investment for all major space acquisitions from $6.31 billion to $9.22 billion, or about 46 percent over the next 3 years. More may be needed if technical, software, and other problems on current programs worsen. At the same time, investment needs for other weapon systems are also on the rise, while long-term budget forecasts indicate that considerably fewer dollars will be available for discretionary spending in coming years rather than more. Funding for Space Radar is further complicated by the lack of long-term funding agreements beyond fiscal year 2013, adding uncertainty to DOD's and the intelligence community's ability to afford expensive programs such as Space Radar. To its credit, Space Radar has worked to establish a key funding agreement between DOD and the intelligence community that addresses short-term cost sharing responsibilities. In prior reports, we have stated that as long as too many programs compete for too few dollars in DOD, programs will be incentivized to produce optimistic estimates and suppress bad news. They will view success as securing the next installment of funds versus delivering capability within cost and schedule goals. We have recommended that DOD guide its decisions to start space and other weapons acquisition programs with an overall investment strategy that would identify priorities for funding so that space systems that are expected to play a critical role in transformation, such as Space Radar and TSAT, could be priorities along with other legacy and transformational systems. To date, this has not been done for space or for DOD's broader weapons portfolio.
- Schedule: Schedules for both programs may also be optimistic. The TSAT program may have underestimated the time for design, integration, and production activities. For example, TSAT embarked on a major software development effort in January 2006 that would build the overall network architecture and provide network management capabilities for TSAT and Advanced Extremely High Frequency satellites, but DOD's Program Analysis & Evaluation office has expressed concern about the overall complexity of the program and the ability of the contractors to write enough software code in one year as is necessary for the program to proceed effectively. In addition, the Space Radar schedule is shorter between program start and initial launch capability than what DOD has achieved for other complex satellite systems. The Space Radar acquisition timeframe from program start to initial launch capability is 86 months, which our analysis shows is shorter than what DOD has achieved or estimated for other complex satellite systems.
- Workforce: TSAT also faces further challenges in meeting workforce personnel requirements to manage and oversee the program in the future, such as the impact from future Air Force workforce reductions of 40,000 active duty personnel—positions that the Air Force may not be able to fill with civilians due to budgetary constraints.

CONCLUSION

Continued efforts by the programs to instill best practices on TSAT and Space Radar are good steps toward addressing acquisition problems, representing significant shifts in thinking about how space systems should be developed. While these steps can help better position these programs for success, they will not work without adhering to commitments to delay milestone decisions or make trade-offs if there are still gaps between requirements and resources. DOD space program and senior officials recognize this and have expressed a commitment to delay program milestones in order to provide the time needed to match resources to requirements, if necessary. However, DOD has not addressed funding pressures that have encouraged premature program starts and too much optimism for past satellite development efforts.

RECOMMENDATION FOR EXECUTIVE ACTION

To ensure that TSAT and Space Radar do not succumb to funding pressures within DOD, we recommend that the Secretary of Defense direct the Under Secretary of the Air Force to identify potential gaps between requirements and resources before approving the start of product development and, if necessary, adjust requirements and resources to increase the likelihood of achieving program cost, schedule, and performance goals.

We provided a draft of this letter to DOD for review and comment. DOD concurred with our recommendation and provided technical comments, which were incorporated where appropriate. DOD's letter is reprinted as Appendix I.

SCOPE AND METHODOLOGY

To assess DOD's progress in adopting best practices as both of these programs proceed toward product development, we obtained and analyzed pertinent documents from the program offices at the Air Force Space and Missile Systems Center at Los Angeles Air Force Base, California. We reviewed budget documents, risk management plans, and risk handling plans as well as requirements documentation for both TSAT and Space Radar. We also reviewed acquisition strategies, program office and prime contractor schedules, and technology development plans for both programs.

To accomplish our work, we conducted interviews with cognizant and responsible program officials at Space and Missile Systems Center in El Segundo, California, and with Department of Defense officials in Arlington, Virginia. We also met with Air Force Space Command officials at Peterson Air Force Base, Colorado, as well as the Space Radar Integrated Program Office in Chantilly, Virginia. We also visited contractor facilities in California, Colorado, and Maryland.

We conducted our work from July 2006 to March 2007 in accordance with generally accepted government auditing standards.

We will send copies of the letter to Department of Defense and interested congressional committees. We will also make copies available to others upon request.

Should you or your staff have any questions on matters discussed in this report, please contact me at (202) 512-4841 or chaplainc@gao.gov contact points for our Offices of Congressional Relations and Public Affairs may be found on the last page of this report. Principal contributors to this report were Art Gallegos, Assistant Director; Josie Sigl; Ann Hobson; Arturo Holguin; Jeff Barron; Rich Horiuchi; Maria Durant; Jackie Wade; Tony Beckham; and Hai Tran.

Cristina Chaplain
Director
Acquisition and Sourcing Management

ENCLOSURE I: COMMENTS FROM THE DEPARTMENT OF DEFENSE

OFFICE OF THE ASSISTANT SECRETARY OF DEFENSE
6000 DEFENSE PENTAGON
WASHINGTON, DC 20301-6000

NETWORKS AND
INFORMATION INTEGRATION

JUL 20 2007

Ms. Christina Chaplain
Director, Acquisition and Sourcing Management
U. S. Government Accountability Office
441 G Street, N.W., Washington, D.C. 20548

Dear Ms. Chaplain,

This is the Department of Defense (DoD) response to the Government Accountability Office (GAO) draft report 07-1029R, 'DOD is Making Progress in Adopting Best Practices for the Transformational Satellite Communications System and Space Radar but Still Faces Challenges,' dated June 19, 2007, (GAO Code 120647). The GAO assessment of the Transformational Satellite Communications System and Space Radar programs was informative and provided additional insight into issues the Department was addressing with the Air Force since early 2006. The Department concurs with the GAO recommendation and enclosed is a response.

The principle action officer for this effort is Mr. Frank Myers. He can be contacted at (703) 607-0289 or by email at frank.myers@osd.mil.

Dr. Ronald Jost
Deputy Assistant Secretary of Defense
(C3, Space and Spectrum)

Enclosure:
As stated

GAO DRAFT REPORT JUNE 19, 2007
GAO-07-1029R (GAO CODE 120647)

"DOD IS MAKING PROGRESS IN ADOPTING BEST PRSCTICES FOR THE TRANSFORMATIONAL SATELLITE COMMUNICATIONS SYSTEM AND SPACE RADAR BUT STILL FACES CHALLENGES"

DEPARTMENT OF DEFENSE COMMENTS TO THE GAO RECOMMENDATIONS

RECOMMENDATION 1: The GAO recommends that the Secretary of Defense direct the Under Secretary of the Air Force to identify potential gaps between requirements and resources before approving the start product development, and if necessary, adjust requirements and resources to increase the likelihood of achieving cost, schedule, and performance goals. (Page 8/GAO Draft Report)

DOD RESPONSE: The Department of Defense (DoD) concurs with the GAO recommendation. DoD agrees that requirements and resources need to be synchronized to ensure space acquisition programs succeed. For space programs, DoD, and specifically the Under Secretary of the Air Force, implemented a Back to Basics philosophy that is focused on maturing technology prior to acquisition and delivery capability in smaller but value-added increments through the use of a Block Approach. In this paradigm, each specific capability increment is based on a balance of capability, delivery timeline, technology maturity, risk and budget. Tradeoffs in these areas mitigate any disconnects between requirements and resources as well between requirements and technology. A Block Approach for space acquisition, coupled with a robust Science and Technology program to mature technologies and reduce risk, greatly minimizes the potential for programs to experience significant cost growth and schedule delays.

ENCLOSURE II: SPACE RADAR BRIEFING SLIDES

Space Radar

Briefing to Staff of the Subcommittees on Strategic Forces Armed Services Committees

Preliminary Findings

March 13, 2007

1

Briefing Contents

- Background
- Objective
- Preliminary Findings
- Conclusions
- Scope of Work
- Back-Up Slides

BACKGROUND

System Description and Capabilities

- Through an integrated program office, the Department of Defense (DOD) and the intelligence community (IC) are collaborating to develop a single common radar system, called Space Radar (SR), to provide global, persistent, all-weather, day and night, intelligence, surveillance and reconnaissance capabilities, particularly in denied areas.
- As envisioned, SR is to consist of a constellation of low earth orbiting satellites, ground systems and communications network, and would generate large volumes of radar data for transmission to ground-, air-, ship-, and space-based platforms.

BACKGROUND

System Description and Capabilities (cont.)

- The core capabilities of SR are to include:
  - Synthetic aperture radar imaging, surface moving target indication, open ocean surveillance, high-resolution terrain information, and advanced geospatial intelligence.
  - Processing, disseminating, and exploiting collected data to support both national and theater users.

4

BACKGROUND

Program Cost and Complexity

- SR could be one of the more expensive and complex space systems DOD has ever tried to develop. According to the program office, system capabilities of the SR constellation will exceed that of any current on-orbit system.
- The Integrated Program Office estimates the cost of developing, producing, and operating the system through 2027 from $20 billion to $25 billion.
- Ground segment processing systems will have to handle the large volumes of data to be produced by the satellites. The program office estimates that the SR ground segment development effort represents one of the most significant challenges to the program and may involve about 5.3 million lines of new and reused software code.

5

BACKGROUND

Management and Stakeholders

- Through the Integrated Program Office in Chantilly, VA, the Air Force, National Reconnaissance Organization (NRO), and National Geospatial-Intelligence Agency (NGA) are responsible for space and ground segment development.

- The primary stakeholders are those agencies who will be developing, operating, supporting, and using the products of the SR system to support military warfighting and national intelligence requirements as well as civil objectives, including the military services, combatant commands, combat support agencies, Joint Chiefs of Staff, the IC, and civil agencies.

BACKGROUND

Program Status

- The development effort is currently in the concept development phase, focusing on technology development and systems engineering activities.
- Product development is scheduled to begin in fiscal year 2009 and the first satellite is scheduled to be ready for launch in fiscal year 2016.
- With recent congressional concerns and funding reductions, the Under Secretary of the Air Force has re-focused the SR acquisition approach.
- Currently 10 satellites are to be developed (9 plus 1 spare however, the definitive number of satellites is still under consideration until key decision point-B (KDP-B), also known as program start.

BACKGROUND

Knowledge About Requirements and Resources Should Influence Program Start

- Our Best Practices reports show that gaining knowledge about requirements and resources before product development is important for space acquisition success.

- The following steps should occur before acquisition programs are initiated:
  - Fully define and stabilize requirements;
  - Assure other resources will be available (funding, technology, time); and
  - Mature technologies to the point of being tested in a relevant or realistic environment (technology readiness level 6-7) to reduce the likelihood of costly and time-consuming rework during acquisition.

8

Objective

- Assess DOD's efforts to gain knowledge of requirements and resources as the Space Radar development efforts proceed toward product development.

9

Results in Brief

- Program has strived to close knowledge gaps:
  - Requirements: program has developed tools to get agreement and collaboration among users and development partners.
  - Resources:
    - Program is following incremental acquisition approach which focuses on use of mature technologies and is taking other actions to reduce technical risk.
    - Program has reached agreements, reflected in budget, for cost sharing over the Future Years Defense Program (FYDP).
- Challenges remain:
  - Requirements: Key performance parameters still to be defined.
  - Resources:
    - Program may not have planned enough time for design, integration, and production activities.
    - High-level agreements between DOD and the intelligence community for long term (beyond FYDP) cost sharing and defining management roles and responsibilities have not been finalized.
    - Growth in DOD's space investment portfolio raises questions about its ability to afford expensive development efforts such as SR.

10

PRELIMINARY FINDINGS

Program Has Strived to Close Gaps:
Requirements

- Program has developed tools to get agreement and collaboration among users and development partners—at both high and lower levels of management.
  - Requirements and Capabilities Working Group
  - Requirements and Capabilities Group
  - Executive Committee
  - Executive Steering Group
  - Joint Requirements Oversight Council/Mission Requirements Board
- According to the program office, coordination efforts on developing requirements to date have been effective

11

PRELIMINARY FINDINGS

Program Has Strived to Close Gaps: Resources – Technology

- Program is to include mature technologies and not push to adopt advanced technologies (such as on-board processing, lithium-ion batteries, and more efficient solar cells) unless they become mature in time for preliminary design review (PDR). Efforts to develop follow-on satellites have yet to be defined.

12

PRELIMINARY FINDINGS

Technology Expected to Be Mature at Program Start

Technology	Current TRL	Work to Be Done	Expected TRL At KDP-B
Analog to digital converter	TRL 3	Develop space-qualified advanced analog to digital converter.	TRL 5 [1]
Integrated radio frequency assembly	TRL 4	Integration and demonstration of radar tiles and panels (including panel-mounted electronics), radar electronic unit, and front-end processor. Demonstrate an integrated subscale electronically scanned array antenna over simulated expected environments.	TRL 6
Low earth orbit laser communication terminal	TRL 4	Laser terminal to be demonstrated in low earth orbit simulated environment.	TRL 6
Surface moving target indication processing algorithms	TRL 4	Establish and demonstrate algorithm test beds; expand data repository with relevant synthetic/collected data; validate performance against stressing, full-scale datasets.	TRL 6
Open ocean surveillance processing algorithms	TRL 3	Performance of open ocean surveillance processing algorithms to be demonstrated using test bed aircraft, synthetic, and other data to validate performance predictions.	TRL 6

[1] Note: The program office is coordinating plans for demonstrating the maturity of the advanced analog to digital converter. It has established an initial test program but needs to resolve whether or not testing is required at a higher level of assembly to meet the standard for demonstrating technology maturity at KDP-B.

13

PRELIMINARY FINDINGS

Technology Expected to be Mature at Program Start (cont.)

- Current technology readiness levels (TRL) of critical technologies are low–TRL 3 to TRL 4.
- With one exception, the program office expects to have mature critical technologies–TRL 6–at KDP-B (initiation of product development). Concept definition contracts do not stipulate maturing technologies to TRL 6 but require the demonstration of "appropriate" technology maturity for a KDP-B technology maturity assessment.
- Section 2366a of Title 10, United States Code, stipulates that a major defense acquisition program may not receive KDP-B approval until the milestone decision authority certifies that, among other things, the technology in the program has been demonstrated in a relevant environment. According to the program office, this requirement can be satisfied with TRL 6.

14

PRELIMINARY FINDINGS

Program Has Strived to Close Gaps: Resources – Technology

- Technical risk:
  - Program has proactively identified and categorized risks and developed plans to address them.
  - Program established a software division at the same level as other major divisions within the program office to elevate the visibility of software development and oversight.

15

PRELIMINARY FINDINGS

Program Has Strived to Close Gaps: Resources – Cost Sharing

- According to the program office, a short-term agreement through fiscal year 2013 has been established and used for developing the fiscal year 2008 budget estimates. NRO Military Intelligence Program is to fund the SR program at least through FY13 (Air Force provided funding prior to FY08).

16

PRELIMINARY FINDINGS

Challenges Remain – Requirements

- Key performance parameters still to be defined and most requirement performance specifications remain to be finalized or determined.

17

PRELIMINARY FINDINGS

Challenges Remain – Program May Not Have Planned Enough Time

- GAO analysis shows acquisition time-frame from program start (KDP B) to initial launch capability (ILC) for SR is shorter than what DOD has achieved or estimated for other complex satellite systems.
- GAO analysis also shows the time period between preliminary design review (PDR) and critical design review (CDR) for SR is shorter than other major space programs.
- PDR determines whether preliminary designs are complete and if the program is prepared to start detailed design and test procedure development. CDR assesses the systems final design and according to GAO best practices, at least 90% of engineering drawings should be completed to provide tangible evidence that the design is stable.

18

PRELIMINARY FINDINGS

Comparison of Months Between Program Milestones

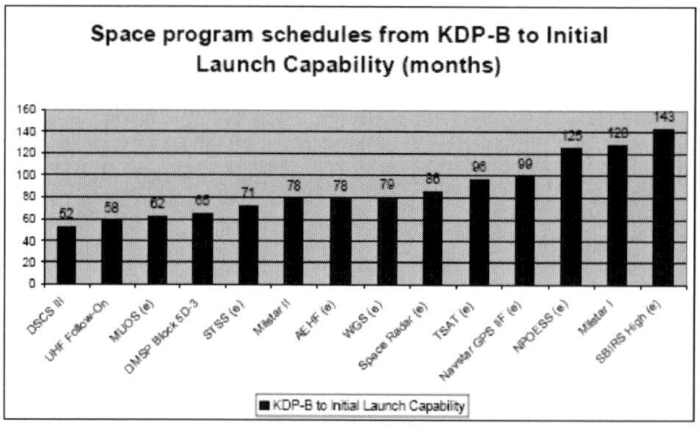

Note: All programs with *(e)* denotation used current estimated dates for Initial Launch Capability.

Source: GAO analysis of DOD data

LEGEND

DSCS – Defense Satellite Communications System

UHF – Ultra High Frequency

MUOS – Mobile User Objective System

DMSP – Defense Meteorological Satellites Program

STSS – Space Tracking and Surveillance System

AEHF – Advanced Extremely High Frequency

WGS – Wideband Global Satellite Communications

TSAT – Transformation Satellite Communications System

GPS – Global Positioning System

NPOESS – National Polar-orbiting Operational Environmental Satellite System

SBIRS – Space Based Infrared System

19

PRELIMINARY FINDINGS

Comparison of Months Between Program Milestones (cont.)

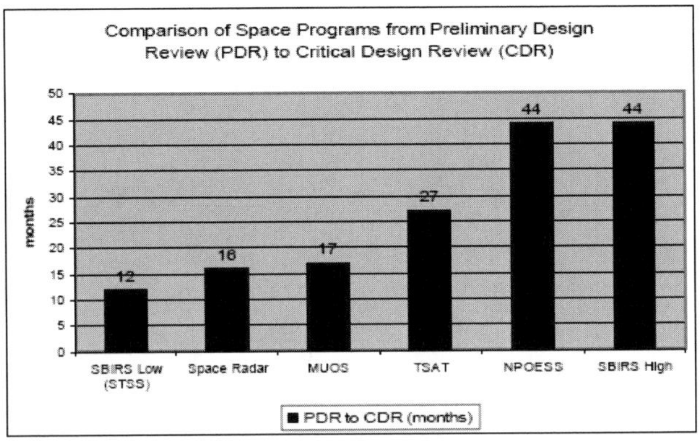

LEGEND

SBIRS – Space Based Infrared System

STSS – Space Tracking and Surveillance System

MUOS – Mobile User Objective System

NPOESS – National Polar-orbiting Operational Environmental Satellite System

TSAT – Transformation Satellite Communications System

Source: GAO analysis of DOD data

20

PRELIMINARY FINDINGS

Challenges Remain – Program May Not Have Planned Enough Time

- Program office officials believe the timeframe is conservative because unlike other programs, they are conducting extensive up-front systems engineering, technology development efforts, and requirements analyses during Phase A. They also stated that the initial launch date is established through systems engineering analyses and a funding availability assessment.
- We agree that the Phase A efforts are reducing risk and have been shown to reduce development time when employed by successful organizations.
- However, the SR development effort is perhaps more complex than other space system acquisitions when considering software and other development activities. E.g., software effort alone expected to be among most complex to date.

21

PRELIMINARY FINDINGS

Challenges Remain: Key risks need to be mitigated

- Program has rated 7 risks as high, including risks related to spectrum, software, and integration with space radar users. Program office acknowledges that some of these risks can not be fully mitigated prior to KDP-B.

Likelihood of Occurrence						
	Near certainty					1
	Highly likely				1	4
	Possible					1
	Unlikely					
	Remote					
		Low	Medium-low	Medium	Medium-high	High
	Severity of Consequence					

Legend: Low risk Medium Risk High risk

Source: Space Radar Integrated Program Office

PRELIMINARY FINDINGS

Examples of top risks of the development effort

Risk and Consequence of Occurrence
If: Program office does not take the necessary steps to understand the periodicity and magnitude of interference on other spectrum users and communicate this risk to prime contractors/payload developer. **Then:** Radar payload performance and mission utility will be reduced.
If: Program office does not take the necessary steps to understand the periodicity and magnitude of interference from other spectrum users and communicate this risk to prime contractors/payload developer. **Then:** Mission performance will be inadequate.
If: Horizontally integrated tasking, processing, exploitation, and dissemination end-to-end requirements are not defined and allocated to Space Radar partner agencies 6 months prior to system design review. **Then:** Horizontal integration goals may not be achieved resulting in performance degradation or cost growth and schedule slippage due to redesign.
If: Integration and testing of the prototype payload panel reveals characteristics that will not meet Space Radar mission requirements or other impacts. **Then:** Significant Changes to panel design may be required, impacting the payload integration and testing schedule.
If: Program office software acquisition process is inadequate for providing SR functionality that will require extensive development by multiple parties and integration of complex software on a schedule that is known to be aggressive. (According to the program office, software in general and large, complex space system software in particular have a history of not meeting schedule, cost, and functionality requirements.) **Then:** Cost increase and schedule slippage and/or functionality, ranging from minor through Nunn-McCurdy breach to possible program failure may occur.

[a]Due to security classification, not all top risks of the development effort are listed.

Source: SR Integrated Program Office

PRELIMINARY FINDINGS

Challenges Remain: Key Risks Need to be Mitigated

- The program office states that it can adequately address these risks because it has or will have sufficient numbers of systems engineers and detailed risk mitigation plans in place. We acknowledge the program's attention to risk mitigation and efforts to bring on systems engineers but do not have evidence to show how its risk mitigation measures go beyond other acquisitions efforts, which were not successful in addressing similar risks.

24

PRELIMINARY FINDINGS

Challenges Remain – Agreements Need to be Finalized: Cost sharing

- Long-term cost-share agreement (beyond FYDP) between DOD and the intelligence community has not been established.
  - In January 2005, the Secretary of Defense and the Director of Central Intelligence committed to share the cost in developing an SR capability.
  - A formal agreement that includes a time period beyond fiscal year 2013 has yet to be signed.
  - Program office expects a cost-sharing agreement to be signed in Spring of 2007.
- Given recent changes in leadership (e.g., Secretary of Defense, Under Secretary for Intelligence, Director of National Intelligence) and the varied interests and missions of the SR development partners, it is important that commitments to cost sharing be formalized soon.

25

PRELIMINARY FINDINGS

Challenges Remain – Agreements Need to be Finalized: Roles and Responsibilities

- Memorandum of agreement between DOD and IC drafted to define management and oversight roles and responsibilities, including defining the milestone decision authority has not been finalized.
  - According to the DOD officials, the memorandum of agreement is to be finalized within 30 days.

PRELIMINARY FINDINGS

Challenges Remain - Program Affordability

- Growth in DOD's space investment portfolio raises questions about its ability to afford expensive programs such as SR.
  - DOD's investment for all major space acquisitions for space from 2006 through 2009 is expected to increase about 46 percent, from $6.31 billion to $9.22 billion.
  - Space Radar is being undertaken at the same time as other major, costly efforts, including Transformational Satellite Communications System, Global Positioning System III, Alternative Infrared Satellite System.
  - In addition to these new programs, DOD is still addressing cost overruns associated with legacy programs like Space Based Infrared Systems High. Moreover, it is likely that DOD will be pressured to increase funding space protection/control. DOD has not developed an overall investment strategy for its portfolio of space programs or conducted affordability assessments which would help prioritize space radar against other space and non-space investments.

Conclusions

- Our best practices work shows that a knowledge-based process can enable decision makers to be reasonably certain about their programs and make informed investment decisions. SR is working toward closing gaps between requirements and resources and has adopted our recommended practices for negotiating requirements and maturing technologies. However, if gaps remain at product development between requirements and resources, SR must be prepared to conduct trade-off analyses or defer milestones to increase the likelihood of achieving program cost and schedule goals.

28

Scope of Work

- Key documents reviewed:
  - SR requirements documentation (Initial Capabilities Document and current versions of the Capabilities Development Document and Concept of Operations)
  - Risk Management Plan and risk handling plans
  - System Acquisition Strategy
  - SR KDP-A Technology Readiness Assessment
  - Program office and prime contractor schedules and technology development plans
  - National Security Space Acquisition Policy
  - Selected Acquisition Reports for major DOD space acquisitions

29

Scope of Work (cont.)

Locations for interviews and documentation:

- Air Force
  - Space Radar Integrated Program Office, Chantilly, VA and Los Angeles Air Force Base
  - Air Force Space Command, Peterson Air Force Base, CO
  - Air Force National Security Space Office, Fairfax, VA
  - Office of the Under Secretary of the Air Force, Washington, DC
- Other Defense
  - Office of the Secretary of Defense, Program Analysis and Evaluation, Arlington, VA
  - Office of the Joint Chiefs of Staff (J-2 and J-8), Arlington, VA
  - Office of the Under Secretary of Defense for Acquisition, Technology, and Logistics, Arlington, VA

30

Scope of Work (cont.)

- Other Defense (cont.)
  - Office of the Under Secretary of Defense for Intelligence, Arlington, VA
  - Office of the Deputy Under Secretary of Defense for Science and Technology, Arlington, VA
  - Institute for Defense Analyses, Alexandria, VA
  - National Geospatial Intelligence Agency, Chantilly, VA
  - U.S. Strategic Command, Offutt Air Force Base, NE
- SR Contractors
  - Northrop Grumman Electronic Systems, Baltimore, MD
  - Northrop Grumman Space Technology, Redondo Beach, CA
  - Lockheed Martin Space Systems Company, Littleton, CO
- Other
  - Congressional Budget Office

We conducted our work from August 2006 to February 2007 in accordance with generally accepted government auditing standards.

31

Enclosure III: Transformational Satellite Communications System (TSAT)

Transformational Satellite Communications System (TSAT)

Briefing to Congressional Committee Staff

March 13, 2007

Briefing Contents

- Objective
- Background
- GAO Findings
- Conclusions
- Scope and Methodology

2

Background

Importance of TSAT

DOD is transforming its military capabilities. As part of this effort, it plans to:

- achieve information superiority over adversaries, and
- share information seamlessly among disparate weapons systems.

One of the key transformation initiatives is the Global Information Grid (GIG), a collection of programs and initiatives modeled after the Internet that is aimed at building a secure information network for enhanced rapid decision making.

3

Background

Importance of TSAT

The Transformational Satellite Communications System (TSAT), the space-borne element of the GIG, is designed to provide more rapid world-wide secure communications with other systems using radio frequency and laser (lasercom) crosslinks.

- For example, in less than a second, TSAT could disseminate a radar image from a Global Hawk that would take Milstar 12 minutes and the Advanced Extremely High Frequency (AEHF) systems 2 minutes to disseminate.

4

Background

Cost, Funding & Schedule

In January 2004, the program's total life cycle cost was estimated at almost $16 billion. A new cost estimate will not be developed until the program moves into the system development and demonstration phase.

TSAT Annual Funding Profile
For Fiscal Years 2005 to 2011

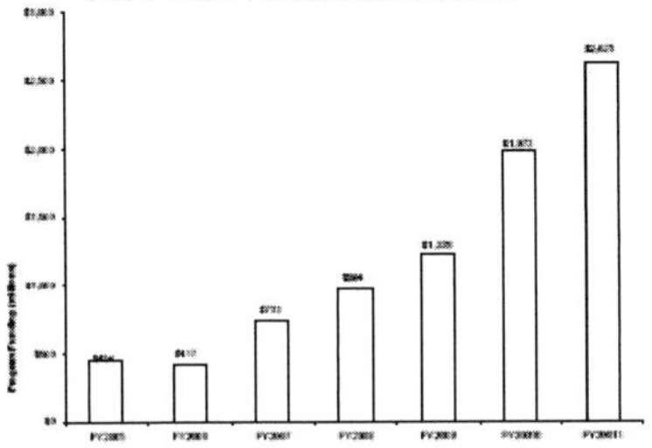

Background

Cost, Funding & Schedule

- The program has spent about $2.1billion since its inception. The funding estimate for FY 07 is almost $733 million.
- Entry into the system development and demonstration phase is scheduled for the first quarter of FY 2008.
- Initial launch is scheduled for first quarter, fiscal year 2016.
- DOD and the Air Force reduced the FY 08 TSAT budget request by about $573 million. The Air Force has also moved the first launch from FY 15 to FY 16.

Background

Acquisition Schedule

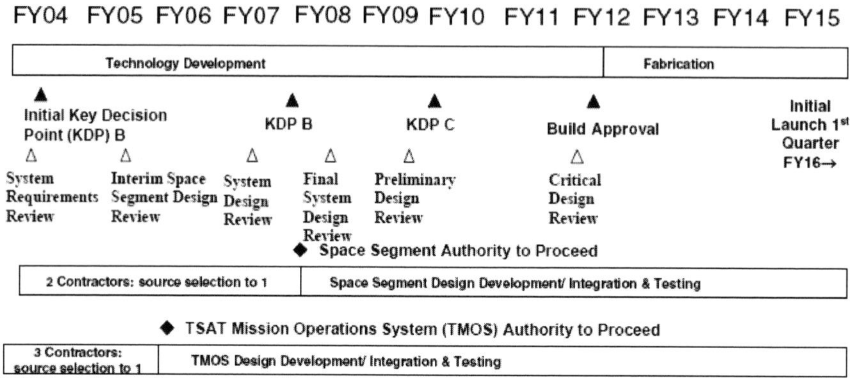

7

Background

Incremental Block Approach

As GAO reported last year, DOD restructured the TSAT program to better position it to gain critical knowledge before it enters the preliminary design phase.

Before TSAT moves to product development:

- all critical technologies must be mature, and
- system design review (SDR) must be complete prior to seeking preliminary design phase approval.

8

Background

Incremental Block Approach

Through a "block" acquisition approach, the program is to:

- control risk through flexibility by scaling the capabilities initially delivered in the first block;
- meet the goal of maintaining scheduled deployment; and
- provide the ability to add additional capabilities in subsequent blocks.

Since the last GAO report, the program has specified what technologies will be included in TSAT Blocks 1 and 2.

9

Objective

Assess DOD's efforts to gain knowledge of requirements and resources as the TSAT program proceeds toward product development.

10

Results in Brief

- TSAT program has continued to gain knowledge about requirements and technologies:
  - Reflected agreements on requirements in specifying future blocks.
  - Made progress in technology maturation activities during FY 2006 and the first quarter of FY 2007, and continues to focus on maturing its key subsystem technologies to a technology readiness level (TRL) 6.
- Challenges remain in matching other resources:
  - Technology:
    - Early tests have revealed challenges in laser communication
    - Limited scalability analyses raises integration risks
  - Time: Program may not have planned enough time for activities involved with networking TSAT to other DOD systems
  - People: Program is not able to fill critical technical positions

11

Preliminary Findings

Program Has Strived to Close Gaps: Requirements

- After restructuring, program worked with users and other stakeholders to reflect agreements on requirements in its plans for the subsequent increment, or block, of TSAT.

- Note on broader DOD requirements: Even with TSAT and other DOD satellites assets, gaps between bandwidth needs and resources are expected to continue to grow, requiring continued dependence on commercial bandwidth. Moreover, systems such as Space Radar, may not be able to rely on TSAT. More data provided in backup slides.

12

Preliminary Findings

Program Has Strived to Close Gaps: Maturing Technology

- TSAT continues to focus on maturing its key subsystem technologies to a technology readiness level (TRL) 6. Systems tested at this level are considered to be sufficiently mature and have been tested in a relevant environment.
- According to the program office, in FY 07, the three technologies not already at TRL-6 remain on track to achieve TRL-6, prior to the preliminary design phase.
- The final test analysis for Phase II testing will not be available until the end of the third quarter of FY 07.

13

Preliminary Findings

Program Has Strived to Close Gaps: Maturing Technology

Critical Technologies	Technology Readiness Level (TRL)	Purpose
Communication-on-the move antenna (COTM)	6	Enables high capacity data communications to small terminals (e.g., one foot antennas).
Packet Processing Payload	6	Converts incoming radio signals into digital data for delivery to the correct Internet-like address..
Information Assurance – Transmission Security	6	Protects transmissions from jamming and interception.
Information Assurance – Space High Assurance Internet Protocol Encryptor (HAIPE)	6	Facilitates security between network nodes.
Bandwidth Efficient Modulation (XDR+)	5	Allows higher capacity protected communications.
Dynamic Bandwidth Resource Allocation (DBRA)	5	Adjusts on-orbit resource allocations more efficiently, which will allow more users to be serviced simultaneously.
Single-access Laser Communications	5	Provides a high bandwidth medium to transmit huge amounts of data between satellites.

14

PRELIMINARY FINDINGS

Challenges Remain: Early tests have revealed challenges in laser communication

Phase I testing involved two major components: Next Generation Processor Router (NGPR) and lasercom. The program office is satisfied with Phase I test results and is proceeding with Phase II tests.

Preliminary Findings

Challenges Remain: Limited Scalability Testing Adds Risk

- As GAO previously reported in 2006, assessing scalability is an integral part of technology development testing.
  - Scalability analysis during the technology development phase can be used to demonstrate whether a satellite can support thousands of users, including those connected via communications-on-the-move technology in a theater of operations.
- According to TSAT officials, scalability analysis to date has been focused on functionality at a very small scale (5-10 users).
- Conducting detailed scalability analysis during the current phase could reduce risk during subsequent integration effort, which is the most risky phase of a satellite program.

Preliminary Findings

Challenges Remain: Program Faces Inherent Integration Risks

- According to an official from the Office of Program Analysis & Evaluation (PA&E), although the TSAT program is making strides in maturing the critical technologies, these new technologies must still be integrated into a single space communications system making the TSAT development effort inherently risky.
- GAO has previously reported on the inherent risks of integration. To ensure program success, it is important that all significant testing issues be resolved before DOD authorizes the program to enter the formal acquisition process.

17

FINDINGS

Software Development Schedule Optimistic

A TMOS cost-plus award fee contract valued at over $2.0 billion was awarded to Lockheed Martin Integrated Systems and Solutions in January 2006 to:

- develop the overall network architecture; and
- provide network management capabilities for TSAT and AEHF satellites.
  - TMOS involves a major software development effort and is to provide, among other things, communications mission planning, policy management, external network coordination, and situational awareness/common operational picture in a secure environment.

18

Preliminary Findings

Challenges Remain: Program may not have planned enough time for networking activities

TMOS delivery schedule may be optimistic.

- PA&E has expressed concern about the overall complexity of the TSAT program and that the TMOS program is optimistic in the amount of software code that can be written in a year.
- DOD and the Air Force reduced the FY 08 TSAT budget request by about $573 million. The Air Force has also moved the first launch from FY 15 to FY 16.

19

PRELIMINARY FINDINGS

Challenges Remain: Program is not able to fill critical technical positions

- Over the next five years, the Air Force is to experience a projected decrease of 40,000 active duty positions. TSAT program expects to be impacted.
  - The program office lacks the authorization to meet its government personnel needs due to the workforce reductions. The program office does not expect to receive the number of personnel requested for FY 2008.
  - Program officials said they will need additional government personnel to carry out oversight and management functions in the long-run.

20

Preliminary Findings

Challenges Remain: Program is not able to fill critical technical positions

Program Office Resources

- From FY 06 to FY 07, the program's budget increased by 77 percent, while program office military and civilian staff increased by 14 percent.
- In addition to the 76 staff for FY 06, the program office currently employs over 100 full-time equivalents from federally funded research development center (FFRDC) (primarily Aerospace Corporation).

21

PRELIMINARY FINDINGS

Challenges Remain: Program is not able to fill critical technical positions

- The program office is currently developing a workforce plan intended to identify the necessary government personnel for the program.
  - Based on prior GAO reports, a workforce plan should include five key elements:
    - involve management, employees, and stakeholders;
    - analyze critical skill and competency gaps between current and future workforce needs;
    - develop strategies to fill identified gaps;
    - build capabilities to address requirements; and
    - monitor and evaluate progress towards achieving strategic goals.

22

Conclusions

- Our best practices work shows that a knowledge-based process can enable decision makers to be reasonably certain about their programs and make informed investment decisions. TSAT is continuing to work toward closing gaps between requirements and resources and has adopted our recommended practices for maturing technologies. However, if gaps remain at product development between requirements and resources, TSAT must be prepared to conduct trade-off analyses or defer milestones to increase the likelihood of achieving program cost and schedule goals.

23

Scope of Work

•Key documents analyzed:

- Risk Management Plan and risk handling plans
- System TSAT requirements documentation (Block Delivery Plan)
- Acquisition Strategy
- Program office and prime contractor schedules and technology development plans
- National Security Space Acquisition Policy
- DOD Funding Estimate Reports

We conducted our work between July 2006 and March 2007 in accordance with generally accepted government auditing standards.

24

In: Military Satellites: Issues, Goals and Challenges
Editor: Abel Chirila, pp. 53-76
ISBN: 978-1-60741-238-0

Chapter 3

Space Based Infrared System High Program and its Alternative*

United States Government Accountability Office

September 12, 2007

The Honorable Bill Nelson
Chairman

The Honorable Jeff Sessions
Ranking Member

Subcommittee on Strategic Forces Committee on Armed Services
United States Senate

The Honorable Ellen O. Tauscher
Chairwoman

The Honorable Terry Everett
Ranking Member
Subcommittee on Strategic Forces Committee on Armed Services
House of Representatives

The Honorable Silvestre Reyes
House of Representatives

The U.S. relies on infrared satellites to provide early warning of enemy missile launches and protect the nation, its military forces, and allies. In 1996, the Department of Defense (DOD) initiated the Space Based Infrared System (SBIRS) program to replace the nation's current missile

* Excerpted from GAO Report GAO-07-1088R.

detection system and provide expanded capabilities to support intelligence, surveillance, and reconnaissance missions. DOD expected to field SBIRS by 2004 at a cost of about $4.2 billion. However, over the past 11 years, SBIRS has proven to be technically challenging and substantially more costly. In an effort to stem cost increases and schedule delays, DOD has restructured the program multiple times, including revising program goals.[1] SBIRS is now estimated to cost over $10.4 billion, and the first satellite launch is expected in 2008. Because of continuing problems with SBIRS, DOD began a parallel alternative effort in 2006 known as the Alternative Infrared Satellite System (AIRSS), to compete with SBIRS and ensure that the nation's missile-warning and defense capabilities are sustained, or possibly provide a follow-on capability to SBIRS. You requested that we assess both SBIRS and AIRSS. As agreed with your office, with respect to SBIRS, we focused on the extent to which DOD is prepared to deliver the first two SBIRS satellites within revised cost, schedule, and performance goals. With respect to AIRSS, we examined the adequacy of DOD's decision to proceed with AIRSS as an alternative to SBIRS as well as whether DOD is attaining the knowledge it needs to position the program for success. To address these objectives, we reviewed schedule and funding information and performed our own analysis of cost and schedule projections using the contractor's 2006 cost performance report data. We also examined the resources committed and planned as well as users' needs for the competing effort. We presented our preliminary findings on SBIRS and AIRSS in briefings to your staffs in March 2007. This letter transmits the information provided in that briefing. We conducted our work between August 2006 and March 2007 in accordance with generally accepted government auditing standards. A copy of the briefing is enclosed.

Results in Brief

Over 12 months after its restructuring, SBIRS still faces challenges in meeting cost, schedule, and performance goals—particularly relating to the development of spacecraft and ground system software. At the time of our review, for example, spacecraft software development efforts were behind schedule by as much as 32 percent. Moreover, management reserves—designed to cover unanticipated work— were being depleted at a much higher rate than anticipated. In addition, DOD has not adequately justified its decision to proceed with AIRSS, and there is disagreement within the department on the purpose and scope of the program. DOD has also not adequately positioned the program for success. For example, a demonstration satellite is not being planned in a way that would maximize DOD's ability to incorporate knowledge gained into the AIRSS program. Based on these findings, we recommend that DOD reexamine the AIRSS program. DOD concurred with our findings and recommendation.

Background

DOD initiated the SBIRS program to meet all military infrared surveillance requirements through a single, integrated system and to provide better and timelier data to the Unified Combatant Commanders, U.S. deployed forces, U.S. military strategists, and U.S. allies. SBIRS is to replace the existing infrared system, the Defense Support Program (DSP), which has provided early missile warning information since the 1970s. The SBIRS program was originally conceived as having high- and low-orbiting space- based components and a ground segment for mission data processing and control to improve current capabilities. However, in 2001, the SBIRS Low component was transferred from the Air Force to the Missile Defense Agency and renamed the Space Tracking and Surveillance System. The Air Force continued developing SBIRS High (herein referred to as "SBIRS"). It, along with its associated ground segment, is one of DOD's highest priority space programs. Originally, SBIRS consisted of four satellites in geosynchronous earth orbit (GEO), two infrared sensors placed on separate host satellites in highly elliptical orbit (HEO)—known as "HEO sensors"—and a ground segment for mission-data processing and control. The Air Force also had planned to acquire a fifth GEO satellite to serve as a spare that would be launched when needed.

Since its inception, SBIRS has been burdened by immature technologies, unclear requirements, unstable funding, underestimated software complexity, poor oversight, and other problems that have resulted in billions of dollars in cost overruns and years in schedule delays. These problems have been documented in GAO reports as well as independent teams chartered by DOD.[2] In addition, the program has been restructured several times to account for cost and schedule problems. The Office of the Under Secretary of Defense for Acquisition, Technology, and Logistics (USD (AT&L)) directed the Air Force to begin parallel efforts to develop a viable competing capability for SBIRS, referred to as AIRSS. USD (AT&L) identified three overall objectives for AIRSS:

- motivate successful execution of the SBIRS program by initiating a viable competing capability;
- pursue an alternative approach, with acceptable technical risk that offers DSP-like missile warning capability to ensure current capabilities are sustained, if SBIRS GEO falters; and
- develop a next-generation SBIRS system to meet worldwide surveillance requirements by initiating efforts for technology risk reduction, system definition, and evaluation of alternative sensor architectures.

In 2006, the Air Force completed studies that provided recommendations for technology development, a road map for inserting those technologies, options for future infrared systems that offer the potential to improve the performance and reduce the cost of SBIRS, and an acquisition strategy for AIRSS. In 2006, program officials awarded contracts that aim to advance key technologies and capabilities. USD (AT&L) directed the Air Force to have the first AIRSS satellite available for launch no later than May 2015. The Air Force has budgeted over $3.3 billion for AIRSS, from fiscal year 2007 through 2013.

Space Based Infrared System

Although the Air Force has acted to reduce risks in the SBIRS program and has had some recent successes, the program still faces risk of not delivering promised capabilities within its revised goals. To reduce risk, the SBIRS program cut back on quantity and capability in the face of escalating costs. It deferred capabilities, such as mobile data processors for the Air Force and the Army and a fully compliant backup mission control facility, and it pushed off a decision to procure the third and fourth satellites. The Air Force also concurrently initiated AIRSS as a secondary means of achieving the same capability. However, about 11 months after the most recent SBIRS program's restructuring, a November 2006 assessment report by the Defense Contract Management Agency (DCMA) showed that some efforts within the program were experiencing significant cost increases and schedule overruns and that the outlook is worsening. Furthermore, the program is rapidly spending its management reserves—funds set aside to address unexpected problems.

DCMA projected additional cost increases in some areas, including satellite sensor and software development as well as satellite integration, test, and assembly. Our analysis of data from the contractor's cost performance reports shows that some of these areas have been particularly problematic and are still exceeding cost and schedule goals, even after the program's recent restructuring. For example, we estimate that the cost of the satellite's pointing and control assembly will increase about $25 million by the time its fabrication is complete. Although the program has put into place strategies for deferring work that can be completed at a later time—thereby reducing cost and schedule risks—some work cannot be deferred. For example, some software development activities for the satellite's pointing and control assembly can be completed at a later date. However, all of the software related to the satellite's flight system must be completed before launch. In addition, DCMA program assessment reports through November 2006 indicated that efforts to develop software are significantly behind schedule. In particular, efforts to develop software for the spacecraft lagged behind schedule—by as much as 32 percent. Furthermore, due to the growing amount of rework resulting from unresolved software discrepancies, DCMA estimated that further schedule slips in software delivery are likely to occur. During our review, program officials acknowledged that some of the problems identified by DCMA were still an issue.

In addition, the contractor's management reserve, which is supposed to last through 2012, has decreased from about $232 million to $166 million (about 28 percent) within a span of about 7 months, indicating that unexpected problems continue to emerge. For example, additional development and testing for spacecraft software and issues with the satellite sensors have necessitated additional expenditures. Our analysis shows that if this expenditure rate continues, an additional $500 million, or more, will be required through September 2012. In addition, the program has deferred needed capabilities to meet cost and schedule goals, but the costs associated with these capabilities are not part of the program's total cost estimate. The cost associated with fully fielding these capabilities is estimated to be about $491 million.

Alternative Infrared Satellite System

In reviewing DOD's decision to pursue AIRSS, we found that USD (AT&L) established objectives for the program that were incompatible given the time frame and budget to complete the work under each objective. One was to solely ensure current missile-warning and defense capabilities are sustained, and the other was to develop the next generation of missile-warning and defense systems. The first would require DOD to pursue a low-risk technology path in order to deliver capability quickly. The second would require DOD to advance technologies and/or design and, thus, budget more time for knowledge building in advance of an acquisition program. In other words, one objective served as an insurance policy for SBIRS; the other was a major effort to advance the way DOD detects missile launches.

Subsequently, USD (AT&L) never clarified what was wanted from the program, and the Air Force, in turn, set out to develop advanced capabilities. Moreover, we found that there was disagreement within OSD as to whether the approach being pursued for AIRSS was the only and/or best option available to the Air Force. For example, DOD's Cost Analysis Improvement Group as well as Program Analysis and Evaluation staffs expressed concern that the focus on developing technology would hinder delivery of an AIRSS satellite available for launch in 2015. During our review, it also became evident that AIRSS could not realistically serve as a back-up to SBIRS because the proposed satellite delivery schedule is very aggressive for meeting the 2015 launch availability date, according to AIRSS program officials.

In addition, in its effort to pursue advanced capability, the Air Force has not positioned the AIRSS program for success. First, in our opinion, not enough time is budgeted for developing and launching the first satellite. At the direction of USD (AT&L), the Air Force set 2015 as the launch date for the first satellite. Our assessment found the period planned between "preliminary design"review and "critical design" review for AIRSS is shorter than for most other major space programs.[3] Specifically, the program is allowing only 12 months from preliminary to critical design review, and 4 years from critical design review to satellite delivery. By contrast, the SBIRS High program took 44 months from preliminary design review to critical design review. Two newer programs, Space Radar and Transformational Satellite Communications System (TSAT), have planned for 16 and 27 months respectively. AIRSS Program officials acknowledged that the current time frame is very optimistic.

Second, the AIRSS program may be optimistic in its assumptions about technology risk. The program's schedule shows critical technologies reaching a high level of maturity at program start, and most are now rated at a technology readiness level of 5 or higher—meaning that the basic components have been integrated with reasonably realistic supporting elements in order that the technology can be tested in a simulated environment. These readiness levels are comparatively higher than other satellite programs we have reviewed. However, we found that the program was still facing considerable technical risk since it is working to build an infrared telescope with a large viewing capability that has never before been developed and it is planning to use "cryocoolers" that have yet to demonstrate low levels of jitter, high efficiency, and long life, and a sensor chip whose assembly's performance level has yet to be verified.[4]

Third, the Air Force's research laboratory officials have stated that on-orbit testing is the only way to validate the proposed capability for AIRSS and reduce risk to an acceptable level. To achieve these results, the Air Force is proposing to launch a small-scale demonstration satellite in late 2010. However, the results from the on-orbit demonstration satellite will not be ready in time to fully inform the development of the first AIRSS satellite. Furthermore, AIRSS officials plan to award contracts for the first satellite before data from on-orbit testing is completed. Our analysis shows that if the tests do not go well, DOD will not have time to return to an approach using lower-risk technology.

CONCLUSION

SBIRS continues to face risks that endanger DOD's ability to sustain, replace, and expand its current missile-warning and defense capabilities. Moreover, the program still has complex and difficult work ahead as it undertakes efforts to integrate technology. Recognizing these risks, the Office of the Secretary of Defense made a sound decision in pursuing the AIRSS program to act as an alternative to the third SBIRS GEO satellite. However, the program has since diverged from this purpose and opted to pursue a higher risk effort in order to advance capability. Moreover, the Air

Force has added risk to this effort by compressing the schedule and limiting the knowledge gained from the demonstration effort. While it is acceptable in any given portfolio to take some high risks, it is not sound for all investments to be high risk—particularly when the capability is as critical to the conduct of military operations as the mission-warning capability is.

Recommendation

We recommend that the Secretary of Defense direct the Under Secretary of Defense for Acquisition, Technology, and Logistics to reassess its investment in AIRSS and alternative ways of reducing the risk posed by the SBIRS program, to more confidentially assure that current missile-warning and defense capabilities are sustained.

Agency Comments and Our Evaluation

We provided draft copies of this letter to DOD for review and comment. DOD concurred with our findings and recommendation. DOD's letter is reprinted as Enclosure I.

Scope and Methodology

To determine the SBIRS program's ability to meet cost and schedule projections, we examined schedule and funding information for developing hardware and software. We reviewed formal program reviews and DCMA and contractor-performance reports. In addition, we performed our own analysis of cost and schedule projections using the contractor's, Lockheed Martin Space Systems Company, 2006 cost performance-report data. To determine the potential problems and risks relating to cost, schedule, and performance that are still facing the SBIRS program, we reviewed technical reports and program briefings and held discussions with program and contractor officials regarding ongoing challenges.

To assess efforts in attaining the knowledge DOD needs before the start of a competing effort, we examined the resources (technology, communications infrastructure, and funding) committed and planned for the competing effort as well as the users' needs for the competing system. We considered DOD's plans for maturing the critical technologies when we obtained technology-readiness information for each critical technology against best-practice standards to determine if technologies will be sufficiently mature when DOD plans to start product development. We also reviewed risk-management plans and concept-development information.

We will send copies of the correspondence to Department of Defense and interested congressional committees. We will also make copies available to others upon request. In addition, the report will be available at no charge on GAO's Web site at http://www.gao.gov.

Should you or your staff have any questions on matters discussed in this correspondence, please contact me at (202) 512-4841 or chaplainc@gao.gov. Contact points for our Offices of Congressional Relations and Public Affairs may be found on the last page of this correspondence. Principal contributors to this report were Arthur Gallegos, Assistant Director; Maricela Cherveny; Claire Cyrnak; Jean Harker; Leslie Pollock; and Greg Campbell.

Sincerely yours,

Cristina Chaplain
Director
Acquisition and Sourcing Management

ENCLOSURE I: COMMENTS FROM THE DEPARTMENT OF DEFENSE

OFFICE OF THE ASSISTANT SECRETARY OF DEFENSE
6000 DEFENSE PENTAGON
WASHINGTON, DC 20301-6000

NETWORKS AND INFORMATION INTEGRATION

August 31, 2007

Ms. Christina T. Chaplain
Director, Acquisition and Sourcing Management
U.S. Government Accountability Office
441 G. Street, NW
Washington DC 20548

Dear Ms Chaplain:

The Department agrees with comments to GAO Report No. 07-1088R, "Space Based Infrared System High Program and its Alternative," August 1, 2007. Consistent with the GAO recommendations, we are reassessing the Alternative Infrared Space System (AIRSS) and are likewise investigating alternative ways of reducing the Space Based Infrared System (SBIRS) risks.

One such example of the commitment to aggressively investigate alternative SBIRS risk reduction strategies is the Under Secretary of Defense for Acquisition, Technology and Logistics (USD(AT&L)) involvement in conducting quarterly program reviews to address cost, schedule, and performance issues with senior leadership across the Department and our industry partners. Additionally, the Acting USD(AT&L) has established a team of software experts under the Director, Defense Research and Evaluation (DDR&E) to assess the software development effort and assist the SBIRS Wing in minimizing programmatic risk.

The Department is also reexamining the AIRSS program to clarify program objectives and ensure our nation's missile warning and defense capabilities are sustained. It is expected these changes will be reflected in the Fiscal Year (FY) 2009 budget that will continue technology maturation for future space-based infrared applications. We look forward to continuing the close relationship the Department has with the GAO on space-based infrared systems.

Sincerely,

John R. Landon
Deputy Assistant Secretary of Defense
(C3ISR & IT Acquisition)

Enclosure II: Briefing Slides

Missile Detection Systems Development

SBIRS/AIRSS Briefing to Congressional Staff
Preliminary Findings
March 13, 2007

1

Briefing Overview

- Due to extreme cost, schedule, and technical problems, DOD has recently restructured the Space Based Infrared System (SBIRS) High missile detection satellite program and undertaken a competing effort, known as the Alternative Infrared Satellite System (AIRSS), in order to ensure continuity of missile detection operations.

- Per requests by the Subcommittee on Strategic Forces, Senate Armed Services Committee and the Subcommittee on Strategic Forces, House Armed Services Committee and discussions with committee staff members, we assessed both SBIRS High and AIRSS.

2

Briefing Overview

• Due to extent and significance of findings, we plan to brief first on SBIRS High, then on AIRSS

- SBIRS High
  - Background
  - Program Status
  - Objective
  - Results in Brief
  - GAO Findings
- AIRSS
  - Background
  - Program Status
  - Objectives
  - Results in Brief
  - GAO Findings
- Conclusions
- Scope and Methodology

3

Background

- SBIRS High was undertaken in 1996 to improve DOD's missile detection/warning capabilities.
- The program has experienced schedule slips of at least 6 years and cost increases that have triggered legislative requirements to reassess and recertify the program several times—most recently in spring 2006.
- While DOD's total program cost estimate was about $4.1 billion, it is now $10.4 billion—more than a 300 percent unit cost increase.
- Our reviews have attributed past problems to an acquisition approach that decreased oversight of contractors, technology challenges, and software development problems.

4

Background: Space Based Infrared Systems

Capabilities

Legacy System: Defense Support Program (DSP)

- Missile Warning (with a classified Probability of Detection) for North America
- Detection and reporting of strategic and theater ballistic missiles and other infrared events of interest

In Development: SBIRS High

- DSP Successor
- Higher Sensitivity—Sees Dimmer Objects, More Often
- Taskable Sensor—Can do Many Missions at Once
- More Accurate Estimate of Missile Location—Including Launch Point and Impact Point

In Development: Space Tracking and Surveillance System

- Space Tracking and Surveillance System formerly called SBIRS Low
- Critical Element of the Ground-based Midcourse Defense — Managed by Missile Defense Agency
- Tracks Threat Missiles Through Entire Flight—From Launch to Intercept
- Sees Extremely Dim Targets—Can Track and Discriminate Objects That are Not Burning
- Inherent Capabilities Also Support Other Missions

MISSIONS	DSP	SBIRS H	STSS
MISSILE WARNING	X	X	o
MISSILE DEFENSE		X	X
TECHNICAL INTEL		X	o
BATTLESPACE CHARACTERIZATION		X	o

X – Primary Mission o – Offers Inherent Capability

5

SBIRS High Program Status: Restructured Acquisition Strategy

- Restructured Acquisition Baseline
  - Up to 3 GEO satellites and 2 highly elliptical orbit (HEO) sensors on classified hosts
  - Milestone Decision Authority will decide whether to procure GEO 3 based on the performance of GEO 1
  - Worldwide system requirements can not be met with 3 GEO satellites--a constellation of 4 satellites is needed
    - A plan for follow-on missile warning satellites is needed
- December 2005 Acquisition Decision Memorandum (ADM)
  - Directs that maintaining schedule, even at the sacrifice of performance, should be the key to program management

6

SBIRS High Program Status: Recent Program Accomplishments

HEO PAYLOAD
- Initial on orbit performance exceeding specifications

GEO PAYLOAD:
- 100% of GEO 1 payload components delivered and integrated
- September 2006: Completed Acoustic Testing

GEO BUS:
- 95% GEO 1 bus components delivered and integrated
- August 2006: Completed Spacecraft Functional Testing
- December 2006: Began First-time Thermal Vacuum Testing

7

Objective: SBIRS High

- **Determine to what extent DOD is prepared to deliver the first two SBIRS satellites within cost, schedule and performance goals.**

8

Findings: SBIRS High

- In brief, we found that SBIRS High continues to face substantial risks as it enters what is perhaps the most difficult phase of development—integration. Even though the program has just recently been restructured, management reserves[1] are being spent at higher than anticipated rates, and cost and schedule variances are deteriorating.
- Faces challenges in meeting cost and schedule goals
- Total program cost not fully accounted for
- Faces challenges in software development

[1]A management reserve budget is an amount of the total allocated budget withheld by contractors for management control purposes.

9

Finding: Challenges to Meet Cost and Schedule Goals

- SBIRS High program not meeting cost and schedule goals
  - Program assessment reports by the Defense Contract Management Agency (DCMA), after restructuring, show that cost and schedule variances are deteriorating
    - November 2006 DCMA report shows Variance at Completion at about $25.6 million; causes include integration test and operations, thermal vacuum test preparations, and engineering rework
  - GAO analysis shows that some efforts that were having trouble before the February 2006 re-baseline, such as assembly, integration, and test, are still over-running cost and schedule, and that contractor estimates are overly optimistic

10

Finding (cont.): Challenges to Meet Cost and Schedule Goals

- Management Reserve (MR) expenditure rate is not sustainable and may result in a need for additional funding
  - Contractor (Lockheed Martin) spent approximately 28 percent ($66 million) of its MR from April 2006 to November 2006
  - MR designed to last to 2012; at current rate contractor MR will deplete by May 2008 (5 months before projected GEO 1 launch date)
  - If this trend continues, $500 million in additional MR will be required through September 2012

11

Finding: Total Program Cost Not Fully Accounted For

- To meet SBIRS validated operational requirements, 2 HEO payloads and 4 GEO satellites are needed to complete the constellation and achieve required coverage
- Current program unit cost for GEOs 1-3 is about $3.5B
- Fourth GEO satellite not part of current acquisition baseline
  - Cost of GEO 4 and/or 5 satellites have been estimated but there are uncertainties due to several factors including industrial base issues, technology obsolescence, number of satellites to be procured, and engineering rework

12

Finding (cont.): Total Program Cost Not Fully Accounted For

- SBIRS High program has deferred needed capabilities to meet cost and schedule goals; these capabilities are not part of the program's total cost estimate
- Additional funds will be needed if these and other deferred capabilities are eventually provided

Capability	FY06-FY11 ($ mil.)	To Completion ($mil.)	Total ($ mil.)
Mission Control Station-Backup Facility	$75	$0	$75
Integrated Training Suite Increment 2	71	7	78
Air Force Multi-Mission Processors	79	15	94
Army Multi-Mission Processors	62	10	72
Relay Ground Station 3rd Antenna	61	5	66
Ground Refresh	105	0	105
Total	**$453**	**$38**	**$491**

13

Finding: Challenges in Software Development

Spacecraft Software Challenges:

- DCMA program assessment reports through November 2006 show software development is behind schedule
  - Most recent report shows schedule variance around -32 percent
  - DCMA high risk threshold for schedule variance is -5 percent
- Pointing and Control Assembly (PCA) software
  - Restructured work to allow viable off-ramp options
  - Off-ramp options may lower risk
- Flight Software System (FSS)
  - All FSS capabilities are needed for launch – no off-ramp possible
  - Increased schedule risk due to delayed qualification
- Tracking algorithms and software not yet completed or demonstrated
- Hundreds of open deficiency reports -- and growing
- According to the DCMA, there is not enough schedule margin to accommodate unforeseen risks

14

Finding (cont.): Challenges in Software Development

Ground Software Challenges:

- The program office is attempting to reduce the length of time it will take to certify the accuracy of the data processed from GEO 1 (currently about two years) by accelerating ground software development
- Late delivery of space software databases to inform the ground software development team has created a lack of coordination between the space and ground systems
- GAO is assessing reported software capability of contractor

15

Alternative Infrared Satellite System (AIRSS)

16

AIRSS Background

- After the most recent Nunn-McCurdy breach, DOD restructured the SBIRS High program and directed the Air Force to begin developing a competing capability to ensure that the nation's missile warning capability is sustained
- The December 2005 Acquisition Decision Memorandum (ADM) also stated the AIRSS effort should:
  - "plan for a new program for space-based Overhead Non-imaging Infrared that generates competition for the SBIRS High GEO 3 satellite"
  - "perform technology risk reduction, perform system definition, and evaluate alternative sensor architectures"
  - "provide insurance against further difficulties encountered on the SBIRS program"
  - "pursue an approach with acceptable technical risk that offers DSP-like missile warning capability"
- There is no full cost estimate of AIRSS because the system has not been defined due to its early stage of development

17

AIRSS Background

- Congress reduced the fiscal year 2007 AIRSS budget by $35 million for program moderation
- The latest program cost estimate from Fiscal Year 2007 through 2013 is over $3.3 billion

	FY 2007	FY 2008	FY 2009	FY 2010	FY 2011	FY 2012	FY 2013	Total FY 2007-2013
Cost ($ in Millions)	67.552	230.887	354.308	543.534	708.657	722.287	736.998	3364.223

Source: Air Force

18

AIRSS Status: Activities since Acquisition Decision Memorandum

- DOD completed 30-day and 120-day studies into alternative infrared concepts (including Wide Field of View (WFOV))
- Space and Missile Systems Center (SMC)/Air Force Research Laboratory issued two Broad Agency Announcement contracts for development of WFOV technologies to Raytheon ($54.4 million) and SAIC ($26 million)
- SMC issued two Program Research and Development Announcement contracts for system design to General Dynamics ($23.3 million) and Northrop Grumman Space Technology ($24.8 million)
- Air Force Space Command drafted a requirements document for global space infrared surveillance systems, a family of systems that includes AIRSS as the SBIRS High follow-on
  - review by the Joint Requirements Oversight Council is scheduled for summer 2007
- National Security Space Office (NSSO) completed its infrared architecture study in December 2006

19

Objectives: AIRSS

- **Determine how DOD has justified its decision to proceed with preliminary efforts for an alternative to SBIRS High, known as the Alternative Infrared Satellite System (AIRSS)**
- **Determine to what extent DOD is taking the steps necessary to build the knowledge it needs to pursue AIRSS**

20

Findings: AIRSS

- Specifically, AIRSS is not being pursued as a "plan B" program as originally envisioned. Rather than seek to maintain continuity of operations, the program is focused on advancing capabilities. Moreover, it is doing so within highly compressed timeframes. There is disagreement among DOD stakeholders as to the wisdom of this approach, given past experiences with space acquisitions.
- The Air Force is moving forward and pursuing two potentially incompatible objectives within one program:
  1) ensure sustainment of current missile warning capability, and
  2) develop next-generation SBIRS High
- Faces challenges in meeting its target delivery date of 2015

21

Finding: Air Force pursuing incompatible goals within AIRSS

- The December 2005 ADM directs the Air Force to:
  - develop a competitor for GEO 3 to ensure sustainment of missile warning capability (DSP-like)
  - pursue technology risk reduction
- However, the AIRSS program is primarily focused on technology that is newer than the technology on SBIRS GEO
  - FY07 funding is largely being spent pursuing these technologies that pose technical challenges
- The AIRSS on-orbit demonstration satellite results will not be ready in time to fully inform development of the first AIRSS satellite

22

Finding (cont.): Air Force pursuing incompatible goals within AIRSS

- Some DOD officials believe AIRSS should be following lower-risk approach in order to assure continuity of missile detection capability
  - Cost Analysis Improvement Group and Program Analysis and Evaluation staff expressed concern that the focus on technology development will hinder the delivery of an AIRSS satellite available for launch in 2015
- According to AIRSS officials, the Air Force wants to look to the future promised by current technologies, not spend time resurrecting DSP-like technologies. They assert that there are no viable alternatives beyond these two paths. However, others in DOD disagree and believe the 120-day study should have been more robust.

23

Finding: AIRSS may face challenges in meeting launch date of FY 2015

- AIRSS program technology schedule shows critical technologies reaching a high level of maturity by program start. Moreover, most are now rated at a technology readiness level (TRL) 5 or higher.
- However, the WFOV capability has not been achieved with the materials and technologies specifically planned for AIRSS. For example,
  - Mercury-cadmium-telluride sensor chip assembly performance level has yet to be verified
  - Full-earth infrared telescope has yet to be developed and demonstrated
  - Cyrocoolers have yet to demonstrate low jitter, high efficiency and long life

24

Finding: AIRSS may face challenges in meeting launch date of FY 2015

- Some DOD stakeholders also believe the WFOV technology development schedule is optimistic.
- Moreover, according to the Air Force Research Laboratory, flight testing is the only way to validate the WFOV approach to a point where risk is significantly lowered. But this test is not scheduled to happen until 2010.
- GAO believes that if the flight test does not go well, DOD will not have time to return to a lower risk technology effort that focuses on sustaining current capability.

25

Finding (cont.): AIRSS may face challenges in meeting launch date of FY 2015

- The AIRSS office assessed current and projected technology readiness levels (TRLs) for enabling technologies for a WFOV capability. Although the technologies offer varying levels of performance, final WFOV design will determine which technologies will be used.

Item	TRL	
	2006	2008
Focal Plane Array Sensor		
•Sensor Chip Assembly – (mercury-cadmium-telluride, or HgCdTe)	5-6	8
•Hybrid Focal Plane Assembly	5-6	8
•Read-out Integrated Circuits	6	8
Cryocooler		
110K (for HgCdTe)	7	8
Spatial filtering algorithms for theater reporting	6-7	7-9
Radiation hardened Field Programmable Gate Arrays (FGPAs)		
•30K gates	8	9
•400K gates	5	6
Processor (300 multimode interference photonic switches (MIPS))	8	9
18° Infrared Telescope	5-6	6
Spacecraft (structure, power, attitude control system (ACS), etc.)	8	8
Communications suite	8	8

26

Finding: AIRSS may face challenges in meeting launch date of FY 2015

- AIRSS officials acknowledge its schedule is "very aggressive" for meeting a 2015 launch availability date
  - 12 months from preliminary to critical design review (CDR)
  - 4 years from CDR to satellite delivery
- AIRSS officials plan to award system contracts before data from key sensor on-orbit testing is completed.
- GAO has reported that it is important for DOD to gain knowledge about technology readiness prior to proceeding with system development.

27

Conclusion

While it is okay in any given portfolio to take some high risks, it is not sound for all investments to be high risk—particularly if the capability being pursued is critical to the conduct of military operations and national security. The SBIRS High program continues to face risks that endanger DOD's ability to sustain its missile detection capability. Recognizing this risk, DOD made a sound decision in pursuing AIRSS to act as an alternative to the third GEO satellite, envisioning this as a low-risk approach. However, the program is now pursuing a higher risk effort than originally envision because it believes there are not other viable alternatives. Given internal disagreements about whether there are other alternatives, DOD should direct the Air Force to further explore the availability of more mature technologies and designs, or justify its decision not to do so.

28

Scope and Methodology

To determine the program's ability to meet cost and schedule projections, we examined schedule and funding information for developing hardware and software. We reviewed formal program reviews, DCMA and contractor performance reports. In addition, we performed our own analysis of cost and schedule projections using Lockheed Martin's 2006 cost performance report data. To determine the problems and potential risks relating to cost, schedule, and performance that are still facing the SBIRS High program, we reviewed technical reports and program briefings and held discussions with program and contractor officials regarding ongoing challenges.

To assess DOD's efforts in attaining the knowledge it needs before the start of a competing effort, we examined the resources (technology, communications infrastructure, and funding) committed and planned for the competing effort as well as the users' needs for the competing system. We considered DOD's plans for maturing the critical technologies when we obtained technology-readiness information for each critical technology against best practice standards to determine if they will be sufficiently mature when DOD plans to start product development. We also reviewed risk management plans and concept development information.

29

Scope and Methodology (cont.)

- Key documents reviewed and analyzed:
  - SBIRS High Acquisition Decision Memorandum
  - SBIRS High Program Review Updates to AT&L
  - SBIRS High Program office and prime contractor schedules and technology development plans
  - SBIRS High Program Assessment Reports – Defense Contract Management Agency, Lockheed Martin Sunnyvale
  - SBIRS High Lockheed Martin Cost Performance Reports
  - AIRSS Office update briefings
  - AIRSS 120-day study report
  - National Security Space Acquisition Policy
  - Selected Acquisition Reports for major DOD space acquisitions
  - DOD budget documentation for SBIRS High and AIRSS

30

Scope and Methodology (cont.)

Locations for interviews and documentation:

- Air Force
  - SBIRS High Program Office and AIRSS office, Los Angeles Air Force Base
  - Air Force Space Command, Peterson Air Force Base, CO
  - Air Force Research Laboratory, Kirtland Air Force Base, NM
  - Office of the Under Secretary of the Air Force, Washington, DC

- Other Defense
  - Office of the Secretary of Defense, Program Analysis and Evaluation, Arlington, VA
  - Office of the Secretary of Defense, Cost Analysis Improvement Group, Arlington, VA
  - Office of the Joint Chiefs of Staff (J-2 and J-8), Arlington, VA
  - Office of the Under Secretary of Defense for Acquisition, Technology, and Logistics, Arlington, VA

31

Scope and Methodology (cont.)

- Other Defense (cont.)
  - National Security Space Office, Fairfax, VA
  - U.S. Strategic Command, Offutt Air Force Base, NE
  - U.S. Naval Research Laboratory, Washington D.C.
  - Defense Contract Management Agency, Alexandria, VA and Sunnyvale, CA
- SBIRS High Contractors
  - Northrop Grumman Space Technology, Azusa, CA
  - Lockheed Martin Space Systems Company, Sunnyvale, CA

We conducted our work from August 2006 to February 2007 in accordance with generally accepted government auditing standards.

32

References

[1] DOD restructured the program, to include setting new cost and schedule goals, in 2002, 2004, and 2005.

[2] GAO, Defense Acquisitions: Despite Restructuring, SBIRS High Program Remains at Risk of Cost and Schedule Overruns, GAO-04-48 (Washington, D.C.: Oct. 31, 2003) and GAO, Defense Acquisitions: Assessments of Selected Major Weapon Programs, GAO-07-406SP (Washington, D.C.: Mar. 30, 2007). Report of the Defense Science Board/Air Force Scientific Advisory Board Joint Task Force on Acquisition of National Security Space Programs, (May 2003) (also referred to as the "Young Panel report") and the July 2004 update to this report. Space-Based Infrared System Independent Review Team, Final Report, February 2002.

[3] Preliminary design review determines whether preliminary designs are complete and if the program is prepared to start detailed design and test procedure development. Critical design review assesses the systems final design, and according to GAO best practices, at least 90-percent of engineering drawings should be completed to provide tangible evidence that the design is stable.

[4] "Cryocoolers" are refrigeration devices used to reach cryogenic temperatures, or very low temperatures (below –238 °F, –150 °C, or 123 K). Cryogenic refrigeration technology for satellites enables the performance of onboard infrared sensors to enhance missile detection, conduct intelligence gathering, and enable space situational awareness.

In: Military Satellites: Issues, Goals and Challenges
Editor: Abel Chirila, pp. 77-96
ISBN: 978-1-60741-238-0

Chapter 4

SPACE ACQUISITIONS: DOD'S GOALS FOR RESOLVING SPACE BASED INFRARED SYSTEM SOFTWARE PROBLEMS ARE AMBITIOUS*

United States Government Accountability Office

WHAT GAO FOUND

To mitigate the SBIRS flight software problems, DOD has assessed various alternatives and developed a way to implement the software redesign and oversee its development. In April 2008, DOD approved the redesign effort, which addressed problems with the original design that affected the timing of stored programs, distribution of control between processors, and failure at the hardware interface level. Six review teams comprised of 70 personnel in all evaluated the designs to ensure the technical solutions, development approach, and readiness of test facilities were adequate. DOD and its contractor are now implementing the simplified architecture, developing new software, and testing elements critical to the integration and test of systems. DOD is also improving its program oversight and better managing the SBIRS development, by acting on the recommendations of an Independent Program Assessment; addressing weaknesses in management responsibility, accountability and organizational structure; and establishing a central execution team.

DOD has estimated that the SBIRS program will be delayed by 15 months and cost $414 million in funding to resolve the flight software problems, but these estimates appear optimistic. For example, confidence levels—based on the program's ability to develop, integrate, and test software in time to meet the schedule goal—have been assessed as low.

Further, the review teams who approved the designs to start coding software report that the program's aggressive schedule is a major challenge because it allows "little margin for error." DOD has also introduced risk by granting waivers to streamline the software development processes to meet the aggressive schedule. These allow the program to deviate from disciplined processes in order to compress the schedule and meet the goal. In addition, some software elements are behind schedule, and thousands of software activities and deliverables remain to be

* Excerpted from GAO Report GAO-08-1073, dated September 2008.

integrated. Delay by these other programs could create unintended consequences for the SBIRS launch goal. If DOD should need additional time or encounter problems beyond what was planned for, more funds will be needed and launch of the first satellite in December 2009 could be jeopardized.

Confidence Level to Produce Software in Time to Meet First Satellite Launch Goal

Confidence level	Contractors	Estimated launch goal
Less than 10 percent	Aerospace Corporation	December 2009
5 percent	Galorath, Inc.	December 2009
50 percent	Lockheed Martin	December 2009

Source: U.S. Air Force (data); GAO (analysis and presentation).

Why GAO Did this Study

In 1996, DOD initiated the Space Based Infrared System (SBIRS) to replace the nation's current missile detection system, and to provide expanded missile warning capability. Since then, SBIRS has been restructured several times to stem cost increases and schedule delays, including revising program goals in 2002, 2004, and 2005. These actions were partly due to the challenges of developing sophisticated technologies and software. In 2007, SBIRS had a major setback when flight software for the first satellite underwent testing and failed, a failure caused by design issues. DOD developed a plan for resolving these issues, and revised its cost and schedule goals. GAO has assessed (1) the approach used to mitigate the problems, and (2) the cost and schedule risks and challenges of that approach. To conduct our work, GAO has contacted, met with, and performed detailed work at numerous DOD and contractor offices; and reviewed technical documents on flight software.

What GAO Recommends

GAO recommends that the Secretary of Defense revise cost and schedule goals commensurate with acceptable risk to increase the confidence of success, and require the contractor to adhere to disciplined software practices as a priority to reduce risk. DOD partially concurred with the first recommendation to revise the cost and schedule estimates, and concurred with the recommendation to prioritize adherence to software practices.

Abbreviations

DOD	Department of Defense
FFRDC	federally funded research and development center
GEO	geosynchronous earth orbit
HEO	highly elliptical orbit
IPA	Independent Program Assessment

OSD	Office of the Secretary of Defense
RFSW	reusable flight software
SBIRS	Space Based Infrared System

September 30, 2008 Congressional Committees

In 1996, the Department of Defense (DOD) initiated the Space Based Infrared System (SBIRS), a satellite missile warning system, to replace the nation's current missile detection system and to provide expanded capabilities to support intelligence, surveillance, and reconnaissance. Since its inception, SBIRS has been burdened by underestimated software and technical complexities, poor oversight, and other problems that have resulted in cost overruns and years in schedule delays. DOD had expected to field SBIRS by 2004 at a cost of $4.2 billion; however, SBIRS is now estimated to cost over $10.4 billion, and the first satellite launch is expected in 2009—a 7-year delay.

In 2006, you requested that we review the SBIRS program. In response, we reported on an array of problems the program was still facing, particularly with respect to software development, the expenditure of management reserves, and deferred requirements.[1] Subsequent to our work, SBIRS experienced another major setback in January 2007 when the flight software for the first satellite underwent testing and failed. The flight software controls and monitors the satellite's health and status and is considered a critical component of the satellite. In April 2007, DOD determined that the software failure was caused by design issues that affected the timing of stored programs, among other problems. DOD also developed a plan for resolving the issues, and associated cost and schedule goals.

Given the importance of flight software to the first SBIRS satellite and its cost and schedule impact on the SBIRS program, we agreed to follow up on our work and assess the software management, development, and mitigation efforts. Specifically, we (1) identified DOD's approach to mitigate the SBIRS flight software problems, and (2) assessed the cost and schedule risks and challenges of that approach.

To conduct our work for this report, we contacted the Office of the Secretary of Defense (OSD), Air Force, and contractor offices. We also conducted detailed work and held discussions with both the Air Force and Lockheed Martin on their efforts to manage, mitigate, and redesign the flight software that is to operate, control, and monitor the satellite's health, status, and safety. We reviewed technical software plans, assessments, analyses, and independent reviews pertaining to the flight software's redesign, and held discussions with key Air Force and contractor officials on various aspects of the flight software development for SBIRS. In addition, we drew from our body of past work on weapon systems acquisitions practices and used disciplined software practices as criteria.[2] We conducted this performance audit from April 2008 to August 2008 in accordance with generally accepted auditing standards. Those standards require that we plan and perform the audit to obtain sufficient, appropriate evidence to provide a reasonable basis for our findings and conclusions based on our audit objectives. We believe that the evidence obtained provides a reasonable basis for our findings and conclusions based on our audit objectives. Appendix I further discusses our scope and methodology.

Results in Brief

DOD has assessed various alternatives for mitigating SBIRS' flight software problems and developed a way forward to implement the program's software redesign and oversee its development. In April 2008, DOD approved the overall software redesign effort which was to address problems with the original design that affected the timing of stored programs, distribution of control between processors, and failure at the hardware interface level. Review teams—comprised of personnel from the Office of the Under Secretary of Defense for Acquisition, Technology, and Logistics; Aerospace Corporation; Lockheed Martin Corporate; Air Force Space and Missiles Systems Center Wing; and Software Engineering Institute—evaluated the designs to ensure the technical solutions, software requirements, development approach, and readiness of the test facilities were of adequate quality. Currently, DOD and the contractor are working to implement the simplified architecture, develop additional software, and test elements critical to the integration and test of systems. DOD has also undertaken several initiatives to improve its program oversight and to help it better manage the development, such as acting on several recommendations identified in an Independent Program Assessment to address weaknesses in management responsibility, accountability, and organizational structure, and establishing a dedicated execution team with a focus on managing the first satellite effort.

DOD has estimated that the SBIRS program will be delayed by 15 months and cost $414 million in funding to resolve the flight software problems, but these estimates appear too optimistic. For example, the productivity estimates that are based on the program's ability to develop, integrate, and test software in time to meet the schedule have been assessed as low—by technical contractors—ranging from 5 to 50 percent in confidence for meeting the schedule goal. Further, the review teams who approved the designs to start coding software report that the program's aggressive schedule is a major challenge because it allows "little margin for error." In addition, DOD has introduced program risk by requesting and receiving waivers for the purpose of streamlining important software development processes to meet the aggressive schedule. The waivers will allow the program to deviate from disciplined processes in order to compress the schedule and meet the goal. Finally, some program elements are already behind schedule, and thousands of software activities and deliverables remain that must be integrated without significant consequence across the broad spectrum of development elements, such as integration with ground, space, and database systems. Also, the launch range needed by SBIRS to launch the first satellite is scheduled for use by other satellite programs prior to SBIRS. Delay in these other satellite programs could create unintended consequences. Should DOD need additional time or encounter problems beyond what was marginally planned for, more funds will be needed and launch of the satellite in December 2009 could be in jeopardy.

We are making recommendations to the Secretary of Defense regarding the development of realistic cost and schedule estimates commensurate with acceptable program risk to increase the confidence of success, and adherence to disciplined software practices. DOD partially concurred with our recommendation to revise the cost and schedule estimates based on more realistic assumptions, and concurred with our recommendation to require the contractor to make adherence to disciplined practices a priority. On the recommendation to develop realistic cost and schedule estimates, DOD stated that the current goals are executable on the basis of available management reserve and schedule margin, as well as additional funds that have been approved by Congress in the event of any unforeseeable problems or delays. DOD further stated it would consider modifying the cost and schedule goals based on the results of an ongoing flight software assessment. While DOD's plan to assess software and its willingness to revise the cost and schedule goals appear plausible, we believe this approach falls well short of a more reasonable approach to revise the estimates based on realistic assumptions to increase the confidence of success. In light of the program's risks, poor performance history, and technical challenges expected during integration, we maintain that developing goals based on realistic assumptions would place DOD in a position to achieve cost and schedule goals with greater confidence.

BACKGROUND

DOD initiated the SBIRS program to meet all military infrared surveillance requirements through a single, integrated system, and to provide better and timelier data to the Unified Combatant Commanders, U.S. deployed forces, U.S. military strategists, and U.S. allies. SBIRS is to replace the existing infrared system, the Defense Support Program, which has provided early missile warning information since the 1970s. The SBIRS program was originally conceived as having high- and low-orbiting space- based components and a ground segment for mission-data processing and control to improve current capabilities. In 2001, the SBIRS Low component was transferred from the Air Force to the Missile Defense Agency and renamed the Space Tracking and Surveillance System. The Air Force continued developing SBIRS High (herein referred to as "SBIRS"). It, along with its associated ground segment, is one of DOD's highest priority space programs.

The SBIRS program originally consisted of four satellites to operate in geosynchronous earth orbit (GEO), plus one spare, an infrared sensor placed on two host satellites in highly elliptical orbit (HEO)—known as "HEO sensors"—and a ground segment for mission-data processing and control.

The SBIRS GEO satellite is designed to support two infrared sensors—a scanning sensor and a staring sensor. The first GEO satellite is commonly referred to as GEO 1. Figure 1 shows the GEO satellite that is to operate in space.

Source: Lockheed Martin Space Systems Company, Sunnyvale, California. © 2007 Lockheed Martin Corporation.

Figure 1: SBIRS Satellite.

As a result of past technical and program difficulties experienced during sensor and satellite development, the SBIRS program has encountered cost and schedule increases. These difficulties have led DOD to restructure the program multiple times, including revising program goals in 2002, 2004, and 2005. For example, in 2002, the program faced serious problems with software and hardware design progress and, in the Conference Report accompanying the National Defense Authorization Act for Fiscal Year 2002, conferees recommended cutting advance procurement funding due to concerns about program developments and the unclear status of the SBIRS program. At that time, the first satellite launch slipped from 2002 to 2006. In late 2005, SBIRS was restructured for a third time which stemmed from a 160 percent increase in estimated unit cost, triggering a fourth Nunn-McCurdy[3] breach, which again postponed the delivery of promised capabilities to the warfighter.

Flight Software

The flight system software is expected to control the GEO satellite's mission critical functions and activities. Unlike other software programs that can be deferred and uploaded to the satellite after launch, the flight software cannot be deferred because it is critical to the satellite's operation and function. The flight software is expected to operate, control, and monitor the GEO satellite's health, status, and safety. Based on the original design, the flight software was to operate on two of four computer processors onboard the satellite and perform important functions and operations, such as telemetry, thermal control, power management, and fault detection activities.[4] Figure 2 shows a simplified diagram of the original flight software design.

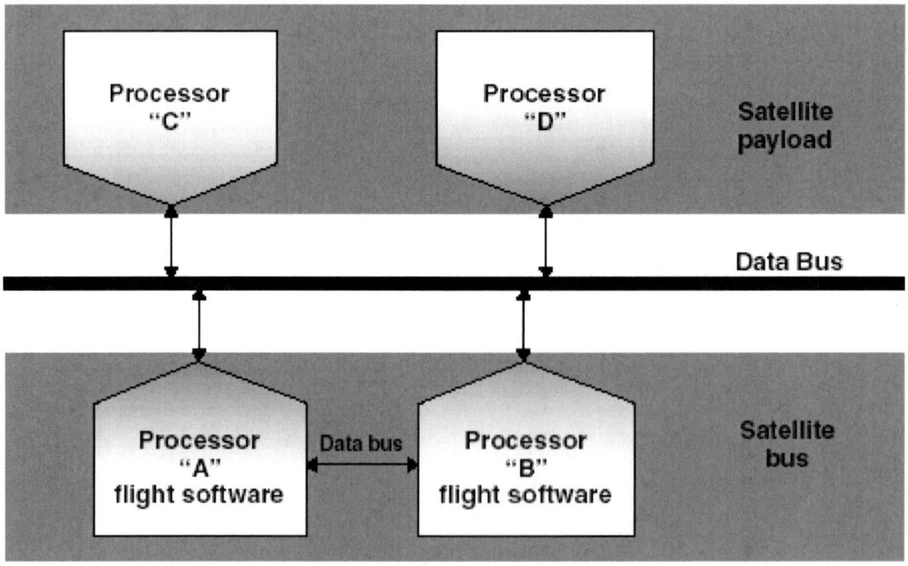

Source: Lockheed Martin (data); GAO (analysis and presentation).

Figure 2. Simplified Diagram of Original Flight Software Design.

Origin and Chronology of Flight Software Events

In 1996, development of the flight software began as an independent research and development project by Lockheed Martin—referred to as reusable flight software (RFSW)—to be used for multifunctional "bus" purposes.[5] In 2004, the RFSW was provided to the SBIRS program for development as the flight system software to operate, control, and monitor the GEO satellite's health, status, and safety. At that time, the software needed to address 1261 requirements in order to satisfy the specific flight software system needs for the GEO satellite. From 2005 to 2006, the Air Force and Lockheed Martin conducted detailed requirements reviews that resulted in the delivery of flight software that was integrated into the satellite's computers.

In January 2007, the flight software underwent testing in a space representative environment called thermal vacuum testing and experienced a higher number of unexpected and unexplained failures. By April 2007, in additional tests, the number of problems escalated well beyond what was expected. At this time, Lockheed Martin notified DOD of the seriousness of the problem. From April 2007 to July 2007, the Air Force and Lockheed Martin analyzed the problems and developed two options:

- modify the existing software or
- redesign the software by simplifying the architecture, developing more software, and increasing the robustness of the fault management system.

The Air Force chose to redesign the software architecture and began its work with Lockheed Martin on detailed software designs from September 2007 to December 2007. In March 2008, the new design underwent Incremental Design Review Block 1 and was approved by the program review board for the revised cost and schedule baseline. In April 2008, six independent review teams examined the Block 2 design during the Systems Engineering & Incremental Design Review and authorized the Air Force and Lockheed to proceed with formal software coding under the redesign.[6]

DOD Is Taking Steps to Mitigate Software Problems, Including Initiatives to Improve Program Oversight

To mitigate the software problems, DOD has assessed various alternatives and developed an approach for implementing the software redesign effort and overseeing its development. DOD and the SBIRS contractor are taking steps to address problems, among others, with the original software architecture. DOD has redesigned the architecture, and is in the midst of developing additional software, and testing elements critical to the integration and test of systems. DOD has also undertaken several initiatives to improve its program oversight and to help it better manage the development, including addressing weaknesses in program management responsibility, accountability, and other areas.

Steps Have Been Undertaken to Address Poor Software Architecture

To address the software's poor architectural design that ultimately resulted in the unexpected loss of telemetry and commanding for extended periods and unexpected hardware errors, a trade study was conducted by Lockheed Martin to examine options for redesign. Table 1 shows the trade study options considered, and recommendations made.

Table 1. Trade Study Options and Recommendations on Software Architecture

Option	Recommendation
Distributed applications (synchronous)	Not recommended due to complexity and risk
Distributed applications (asynchronous)	Not recommended due to complexity and risk; has the highest impact to ground systems
All applications on processor "B"	Not recommended due to complexity and risk
All applications on processor "A"	Recommended as best fit with component and fault management system designs

Source: Lockheed Martin (data); GAO (analysis and presentation).

As indicated in table 1, the trade study recommended a simplified architecture that places all the software applications on a single processor, processor "A", rather than using distributed applications because it represents the best fit with system designs. Lockheed Martin officials stated that the simplified software architecture will address a number of areas that were problematic with the original design, such as the timing of stored programs that failed during thermal vacuum tests. Among other elements, the new design will involve the development of additional software that will also increase the robustness of the fault management system.

Major Redesign Approved for Coding Software

Approved in April 2008, the new designs have undergone numerous reviews, the last of which was subjected to comprehensive and detailed examination involving six independent review teams. Teams comprised of personnel—from the Office of the Under Secretary of Defense for Acquisition, Technology, and Logistics; Aerospace Corporation, a federally funded research and development center (FFRDC)[7]; Lockheed Martin Corporate; Air Force Space and Missiles Systems Center Wing; and the Software Engineering Institute[8]—evaluated the technical solutions, development approach, and readiness of the test facilities, among other elements.

The objective of the design review was to authorize the start of formal software coding. For the incremental design review, independent review teams were provided detailed information about software issues on the original design, including the severity of the issues and the status of each. Other information included DOD's approach in managing risk, resolution of critical issues, disposition of deficiency reports, requirements volatility, and integration with ground systems. Technical data included diagrams of the simplified architecture, operating system interface design, and lines of software code that would be impacted from earlier designs. Other information about the software included designs of subsystems, schematics, integration and delivery schedules, and productivity and sizing estimates.

Progress Is Being Made to Develop Software and Conduct Tests

DOD is making progress to develop needed software and conduct tests of elements that are critical to the first satellite system, called GEO 1. For example, in June 2008, DOD held a design review on software for the fault management system that elicited concurrence from external stakeholders to proceed with coding activities. At the same time, they held a space technical interchange meeting that provided consensus on the methodology and a plan for complete space vehicle testing, including the flight software. In July 2008, Lockheed Martin delivered 63,000 of the projected 67,000 source lines of code for the space vehicle and ground software integration effort, including a database that provided data so that development efforts could continue on ground software and testing activities.

According to Lockheed Martin, software development efforts followed a disciplined process, except in those cases where waivers were requested and granted by the software engineering process group. Figure 2 shows Lockheed Martin's process for developing and qualifying flight software.

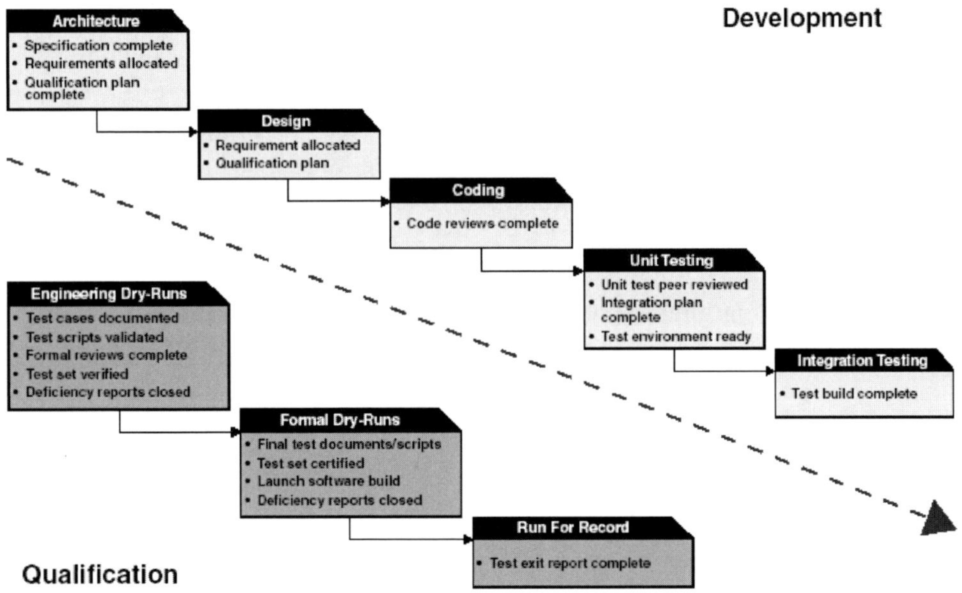

Source: Lockheed Martin (data); GAO (analysis and presentation).

Figure 3. Flight Software Development Process.

Risks Reduced by Funding Additional Test Resources

DOD has taken steps to fund critical test bed resources that are needed to adequately test, model, analyze, and simulate software functions as a means to reduce integration and test risks, in response to lessons learned from the failed software that identified the need to add and upgrade their simulation and test bed resources. For example, an evaluation of the software problems found several contributory factors that prevented them from identifying the software problems earlier. These include:

- test beds that had matured in parallel with the flight software and hardware, making it difficult to distinguish between test bed and software issues;
- oversubscription of test beds and lack of simulation resources that precluded them from checking out high-risk areas (timing, and stored programs); and

- insufficient modeling of timing, and analysis of stored program implementation, which might have shed light earlier on lack of robustness.

In May 2008, the additional test bed and simulator was brought online and is currently in use.

Actions Have Been Undertaken to Address Program Weaknesses, and Improve Oversight of GEO Development

DOD and Lockheed Martin have undertaken several initiatives to address areas of program risk, such as efforts to improve oversight of GEO 1 and flight software development. These include acting on recommendations made in an Independent Program Assessment (IPA) that was conducted to ensure the validity of the technical, cost, and schedule baselines. As part of the assessment, the IPA study assessed contractor performance, evaluated program risk areas, and made recommendations on where program improvements could be made. In November 2007, officials from the Air Force, Lockheed Martin, and Aerospace Corporation reported the IPA findings. Table 2 shows the IPA findings, recommendations, and status of implementation efforts.

Table 2.IPA Findings, Recommendations, and Status of Implementation

Finding	Recommendation	Implemented? (as of April 2008)
1. Lockheed Martin's program process discipline is poor	• Engage Lockheed Martin functional areas and ensure that processes are being followed	Yes
2. Air Force has limited management control over SBIRS	• Amend contract to provide necessary management control	Yes
3. Adversarial relationships exist between Air Force and Lockheed Martin	• Fix responsibility, accountability, and authority disconnects	Yes
4. Government organizational structure is flawed because cost and schedule responsibilities are separated.	• Combine in a single office the review of contractor cost and schedule data	Yes
5. Focal point for FSS completion is needed	• Designate a program manager within flight software system • Establish giver/receiver relationships	Yes

Source: Aerospace Corporation (data) and U.S. Air Force (data); GAO (analysis and presentation)..

As indicated in table 2, the Air Force and Lockheed Martin have taken actions to address areas of risk. Among others, these actions included deliberately emphasizing the software development process where adherence to process disciplines was lacking, and enhancing the interaction between cost and schedule functions where the Air Force organization structure was found to be flawed because it did not mirror the contractor's more traditional approach where these functions are combined for better program control.

To improve the oversight and management of the GEO 1 satellite and software development, the Air Force and Lockheed Martin established a dedicated execution team with a focus on overseeing the test, integration, and assembly of software and hardware, and ensuring delivery of the GEO 1 satellite. The execution team is a joint effort that includes the Air Force, Lockheed Martin, and Aerospace Corporation. As part of the management approach, the execution team is responsible for conducting daily meetings to review "inch stone" metrics and to resolve issues. The execution team also meets weekly with the Executive Program Management leadership to provide early insight on issues and resolve organizational weaknesses, and conduct monthly reviews with senior executives to provide consistent communication and allow opportunity for guidance. According to DOD officials, the execution team not only improved oversight of software development and management of the GEO 1 effort, but also addressed weaknesses identified in the IPA study. For example, these weaknesses included, among others, the need to fix the program's responsibility, accountability, and authority disconnects. Officials reported that the execution team helped alleviate the strained relationships that had existed between the Air Force and Lockheed Martin where adversarial relationships and morale problems were evident.

DOD's Plan for Resolving the Software Problem Is Optimistic

While DOD has estimated that the SBIRS program will be delayed by 15 months and cost $414 million to resolve the software problems, those estimates appear too optimistic, given the cost and schedule risks involved. For example, SBIRS contractors' report low confidence that software can be produced in time to meet the December 2009 satellite launch goal. Further, DOD and the contractor face significant challenges and risks that could result in more time and money being required to meet program goals, to include the bypassing of some disciplined software practices that add risk to cost and schedule. Finally, as of August 2008, DOD reported that SBIRS was already behind schedule on some software development efforts, and thousands of activities remain that must be integrated and tested across various systems, with cost and schedule implications, if problems or unintended consequences occur.

Low Confidence That Software Can Be Produced to Meet Cost and Schedule Goals

A major concern is the infeasibility of producing the software in time to meet the estimated launch goal. For example, technical contractors— Aerospace Corporation, Galorath Inc., and Lockheed Martin—estimated the confidence to be "low" that software can be developed within the tight time frames. These estimates are based on widely accepted models (System Evaluation and Estimation of Resources, Software Estimating Model, and Risk Assessment) that take into account the effective size of the software, staffing of the effort, complexity, volatility of software requirements, and integration and risk of anticipated rework and failure in system tests. Using DOD's self-imposed baseline schedule goal, software productivity estimates show very low confidence levels that the schedule goal can be met. Table 3 shows the confidence in meeting the GEO 1 launch goal in December 2009 (various models used).

Table 3.Confidence Level to Produce Software to Meet GEO 1 Schedule

Confidence level	Contractors	Estimated launch goal
Less than 10 percent	Aerospace Corporation	December 2009
5 percent	Galorath, Inc.	December 2009
50 percent	Lockheed Martin	December 2009

Source: U.S. Air Force (data); GAO (analysis and presentation).

As indicated in table 3, one estimate shows only a 5 percent confidence that the software can be produced in time to meet the schedule goal, while the other estimate shows a less than 10 percent confidence level. Lockheed's own software productivity estimate shows a 50 percent confidence level in meeting the December 2009 launch schedule, but its estimate assumes (1) a higher productivity than has been demonstrated, and (2) the software will require less effort, which has not been the program's experience. According to DOD's Cost Analysis Improvement Group, if productivity on software does not materialize, or problems occur during testing and integration beyond what was marginally planned for, then it could cost an additional $400 million for each year of schedule slippage.

Major Challenge and Risks to the Redesign and Development Effort Still Exist

Based on an April 2008 review of the revised software designs and software development approach, the independent review teams— comprised of personnel from the Office of the Under Secretary of Defense for Acquisition, Technology, and Logistics; Aerospace Corporation; Lockheed Martin Corporate; Air Force Space and Missiles Systems Center Wing; and the Software Engineering Institute—concluded that the program should proceed with formal software coding, but also expressed concern about the ambitious schedule. Specifically, the review teams cited the program's aggressive schedule as a major challenge because it allows "little margin for error" and concluded the program faces high risk of not meeting the schedule. Table 4 shows the weaknesses and risks to software development.

Table 4.Weaknesses and Risks to Software Development

Weaknesses
• Schedule pressure, and alignment of code and designs
• Code complexity impacting unit testing
• Late integration with ground software
• Significant amount of work remaining
Risks
• Concurrent systems engineering and software development
• Code development requiring more labor than estimated
• Additional system or software testing required beyond plans
• Qualification of test products behind schedule
• Systems engineering completion may require more effort

Source: Lockheed Martin (data); GAO (analysis and presentation).

Although the Air Force and Lockheed Martin are committed to the effort and have built in a 120-day margin to fix unexpected and unforeseeable problems, a computer engineer from the Defense Contract Management Agency who is familiar with the program believes that the margin is insufficient because the planned schedule considers only routine development activities, and that additional time will likely be needed to address any unanticipated problems.

Bypassing Disciplined Software Practices Adds Risk

Further, to meet the cost and schedule goals, the program is using approaches that will increase program risk. These risks stem from waivers, which were requested by Lockheed Martin, as specified by software provisions in the program's software development process. In following the SBIRS Software Development Plan, for Flight Software System 1.5, waivers were generated and approved by a software engineering process group so that developers could deviate from the established processes. These deviations from the disciplined development process allowed the program to shortcut important processes in order to meet the ambitious schedule goal, rather than follow a disciplined process to develop software. For example, a waiver was granted for software design to be done in parallel with the software specification activity. However, according to DOD, the risk is that requirements could be rejected and that rework may be required in coding or design. Another waiver was granted for software unit integration testing to be done in parallel with formal unit testing. According to DOD, the risk is that formal unit testing may find problems that were not identified during prior informal (developer) unit testing, thereby necessitating possible rework.

Cost and Schedule Goals Are at Risk Because Some Software Elements Are Behind Schedule, and Complex Integration and Other Activities Remain

Some of the flight software's elements are already behind schedule and a significant amount of activities remain to be done, posing concern to DOD. For example, DOD reported that, as of August 2008, the software qualification test case and script development effort was already a month behind schedule. Also, final delivery of the Block 2 flight software is now forecasted to be at least 2 weeks late. Other problems that could set back SBIRS are the thousands of integration and coordination activities that must take place as they ramp up. For example, Lockheed Martin reports that the schedule has more than 14,500 tasks that will occur, beginning in January 2008, across multiple systems. This means that the flight software test activities and integration efforts must all be integrated in a "single- flow" without consequence across a broad spectrum of systems, such as integration with ground, space, and database systems, among others. Software experts, independent reviewers, and government officials acknowledged that the aggressive schedule, when combined with the significant amount of work that remains, is the biggest challenge facing the program.

Still, there are external factors that could create schedule impacts for meeting the SBIRS schedule goal. For example, DOD reports that the GEO 1 satellite launch could be affected by other satellites scheduled to launch prior to the SBIRS launch. Essentially, these launch activities use the same launch range resources that will be required to launch the GEO 1 satellite, and delays in any of these events could create unintended consequences to the SBIRS GEO 1 launch goal.

Conclusions

Given the technical complexity of the program and SBIRS' poor program history, it is unwise for DOD to pursue such ambitious goals for resolving the flight software problem. More than 12 years after its inception, the SBIRS program continues to face major challenges that have proven technically challenging and substantially more costly than originally envisioned. The testing failure of the flight software is further proof that sophisticated technology and inherent complexities related to software continue to be underestimated. To its credit, DOD has instilled greater discipline by involving outside experts, regaining control of development activities, and dealing with the poor relationships that had existed for some time. To ensure that such steps can lead to success, adherence to disciplined software practices should be made a priority over steps or measures taken to compress the schedule for the sake of meeting the self- imposed launch goal. Prioritizing such disciplines will improve efforts to acquire a better product, increase executability of the program, and reduce program risk. In turn, establishing goals that are synchronized with such priorities will allow DOD to achieve expectations and program deliverables with greater reliability. Essentially, these will position the leadership to better direct investments by establishing goals with greater confidence that they can be achieved.

Recommendations for Executive Action

To better ensure that SBIRS can meet the cost and schedule goals for resolving the flight software problems as well as launch the first satellite on schedule, we recommend that the Secretary of Defense

- revise the cost and schedule estimates based on more realistic assumptions to increase the confidence of success, and
- require that the contractor make adherence to disciplined software practices a priority to reduce program risk.

Agency Comments and Our Evaluation

DOD provided us with written comments on a draft of this report. DOD partially concurred with our recommendation to revise the cost and schedule estimates based on more realistic assumptions, and concurred with our recommendation to require the contractor to make adherence to disciplined practices a priority. DOD's comments appear in appendix II.

In its comments, DOD partially concurred with the recommendation that the cost and schedule estimates be revised based on more realistic assumptions to increase the confidence of success. DOD noted that the current goals are executable on the basis of available management reserve and schedule margin. In the event that the program encounters any unforeseeable problems that may cause further delays, DOD stated that Congress has approved an additional $45 million in funding to mitigate any future launch delays. The department pointed out that OSD is working with the SBIRS program to hold a more specific review of the flight software. Based on the results of this review, DOD stated it would consider them in any decision to modify the cost and schedule estimates. DOD expects these assessments to be complete by the end of the 2008 calendar year.

As indicated in our report, SBIRS has been restructured several times because it underestimated the technical complexity and inherent challenges associated with software, among other technical elements. Neither the software assessment conducted to determine the confidence of producing software nor the independent reviewers who examined the redesign approach indicated that the current goals were executable. Rather, as we noted, software experts, independent reviewers, as well as the government officials we interviewed expressed concern over the aggressive schedule and questionable schedule margin, which the Defense Contract Management Agency believes is insufficient. Moreover, as we previously reported and noted in this report, the expenditure of management reserves has been particularly problematic because these funds were being rapidly spent. Further, while OSD's plan to assess software and its willingness to revise the cost and schedule goals appear plausible, we believe this approach falls well short of a more reasonable approach to increase the confidence of success for the reasons we cited. In light of the program's risks, poor performance history, and technical challenges expected during integration, we maintain that establishing goals that are based on more realistic assumptions would place DOD in a better position to achieve cost and schedule goals with greater confidence.

DOD concurred with the second recommendation stating that adherence to disciplined software development processes improves the quality and predictability of the software development while reducing the amount of rework. DOD further states that the program office and the contractor jointly accepted two process waivers to streamline the process, but that these waivers have had no adverse impact on the software development effort. In order to keep the focus on quality software deliveries, DOD noted that the program would disapprove any waivers which might compromise the team's ability to complete the development.

We are encouraged by DOD's efforts to adhere to disciplined software processes to improve the quality and predictability of development. In this endeavor, DOD states that it would disapprove any waivers that could compromise the development effort. However, it is unclear exactly what criteria DOD will use to determine whether a waiver will compromise development efforts. Without this, there is no mechanism to ensure that any waivers that are granted will not have a material effect on software development.

We also received technical comments from DOD which have been addressed in the report, as appropriate.

We are sending copies of this report to the Secretary of Defense; the Office of the Under Secretary of Defense for Acquisition, Technology and Logistics; the Secretary of the Air Force; and the Director, Office of Management and Budget. Copies will also be made available to others on request. In addition, the report will be made available at no charge on the GAO Web site at http://www.gao.gov.

If you, or your staff, have any questions concerning this report, please contact me at (202) 512-4589. Contact points for our offices of Congressional Relations and Public Affairs may be found on the last page of this report. The major contributors are listed in appendix III.

Cristina T. Chaplain
Director
Acquisition and Sourcing Management

List of Congressional Committees

The Honorable Bill Nelson Chairman
The Honorable Jeff Sessions Ranking Member
Strategic Forces Subcommittee Committee on Armed Services United States Senate
The Honorable Ellen Tauscher Chairwoman
The Honorable Terry Everett Ranking Member
Strategic Forces Subcommittee Committee on Armed Services House of Representatives

APPENDIX I: SCOPE AND METHODOLOGY

To identify the Space Based Infrared System's (SBIRS) approach to mitigate the flight software problems, we reviewed the plans and alternatives the Department of Defense (DOD) put in place to mitigate the software problem. We also interviewed Air Force, Defense Contract Management Agency, and Lockheed Martin officials who were responsible for management and oversight of the software development effort. We also examined technical reports, studies, and analyses about the factors that contributed to the flight software problems, as well as planning documents and alternatives that were considered in fixing the software problem.

To assess the cost and schedule risks and challenges of the way forward, we held discussions with both the DOD and Lockheed Martin on their efforts to assess the program risks and challenges, including their approach to manage, mitigate, and redesign the flight software that is to operate, control and monitor the satellite's health, status, and safety. We also reviewed schedules, risk reports, analyses, program assessments, and independent review reports pertaining to the flight software's redesign, and selected assessments by independent sources that were used, in part, as basis for selecting December 2009 as the launch goal for the GEO 1 satellite. We also interviewed Air Force and contractor officials responsible for developing and executing the redesign, including a contractor hired for their expertise in estimating software productivity.

We conducted this performance audit at the Office of the Secretary of Defense, Washington D.C.; Space and Missile Systems Center, Los Angeles Air Force Base, California; and Lockheed Martin and the Defense Contract Management Agency, Sunnyvale, California from April to August 2008 in accordance with generally accepted auditing standards. Those standards require that we plan and perform the audit to obtain sufficient, appropriate evidence to provide a reasonable basis for our findings and conclusions based on our audit objectives. In addition, we drew from our body of past work on weapon systems acquisition practices and disciplined software practices. We believe that the evidence obtained provides a reasonable basis for our findings and conclusions based on our audit objectives.

APPENDIX II: COMMENTS FROM THE DEPARTMENT OF DEFENSE

OFFICE OF THE UNDER SECRETARY OF DEFENSE
3000 DEFENSE PENTAGON
WASHINGTON, DC 20301-3000

ACQUISITION
TECHNOLOGY
AND LOGISTICS

SEP 29 2008

Ms. Cristina Chaplain
Director, Acquisition and Sourcing Management
U.S. Government Accountability Office
441 G Street, N.W.
Washington, DC 20548

Dear Ms. Chaplain:

This is the Department of Defense (DoD) response to the GAO draft report GAO-08-1073, "SPACE ACQUISITIONS: DoD's Goals for Resolving Space Based Infrared System Software Problems Are Ambitious," dated August 22, 2008, (GAO Code 120761). Detailed comments on the two report recommendations are enclosed.

Sincerely,

Joshua Hartman

Joshua T. Hartman
Director,
Space & Intelligence Capabilities

Enclosure:
As stated

GAO Draft Report Dated August 22, 2008
GAO-08-1073 (GAO CODE 120761)

"SPACE ACQUISITIONS: DOD'S GOALS FOR RESOLVING SPACE BASED INFRARED SYSTEM SOFTWARE PROBLEMS ARE AMBITIOUS"

DEPARTMENT OF DEFENSE COMMENTS TO THE GAO RECOMMENDATIONS

RECOMMENDATION 1: The GAO recommends that the Secretary of Defense, revise the cost and schedule estimates based on more realistic assumptions to increase the confidence of success. (Page 17/GAO Draft Report)

DOD RESPONSE: Partially concur. While the current contractor cost and schedule baseline is aggressive and contains risks, we believe the remaining Flight Software Subsystem (FSS) development is still executable within the available management reserve and schedule margin. Congress has approved an additional $45M in Omnibus funding to provide mitigation against a future launch date delay as a result of any unforeseen program problems, to include Flight Software development and qualification delays. The Wing has recently completed an integrated baseline review of the program. In addition, the program office is working with OSD to hold a more specific review of the FSS effort. These results will be considered in any decision to modify FSS cost and schedule estimates. These assessments will be complete by the end of the calendar year.

RECOMMENDATION 2: The GAO recommends that the Secretary of Defense require that the contractor make adherence to disciplined software practices a priority to reduce program risk. (Page 17/GAO Draft Report)

DOD RESPONSE: Concur. The DoD agrees adherence to disciplined software development processes, as outlined in the SBIRS Software Development Plan, improves the quality and predictability of the software development while reducing the amount of rework. To date, the program office has jointly, with the contractor, accepted the two minor process waivers mentioned in the report to streamline the development process and reduce the schedule risk associated with the December 2009 projected launch date. Those waivers have had no adverse impacts to the FSS development. To keep focus on quality software deliveries in support of space vehicle testing and operations, the program office will disapprove any waivers which compromise the team's ability to complete the development.

GAO Draft Report Dated August 22, 2008
GAO-08-1073 (GAO CODE 120761)

"SPACE ACQUISITIONS: DOD'S GOALS FOR RESOLVING SPACE BASED INFRARED SYSTEM SOFTWARE PROBLEMS ARE AMBITIOUS"

DEPARTMENT OF DEFENSE COMMENTS TO THE GAO RECOMMENDATIONS

RECOMMENDATION 1: The GAO recommends that the Secretary of Defense, revise the cost and schedule estimates based on more realistic assumptions to increase the confidence of success. (Page 17/GAO Draft Report)

DOD RESPONSE: Partially concur. While the current contractor cost and schedule baseline is aggressive and contains risks, we believe the remaining Flight Software Subsystem (FSS) development is still executable within the available management reserve and schedule margin. Congress has approved an additional $45M in Omnibus funding to provide mitigation against a future launch date delay as a result of any unforeseen program problems, to include Flight Software development and qualification delays. The Wing has recently completed an integrated baseline review of the program. In addition, the program office is working with OSD to hold a more specific review of the FSS effort. These results will be considered in any decision to modify FSS cost and schedule estimates. These assessments will be complete by the end of the calendar year.

RECOMMENDATION 2: The GAO recommends that the Secretary of Defense require that the contractor make adherence to disciplined software practices a priority to reduce program risk. (Page 17/GAO Draft Report)

DOD RESPONSE: Concur. The DoD agrees adherence to disciplined software development processes, as outlined in the SBIRS Software Development Plan, improves the quality and predictability of the software development while reducing the amount of rework. To date, the program office has jointly, with the contractor, accepted the two minor process waivers mentioned in the report to streamline the development process and reduce the schedule risk associated with the December 2009 projected launch date. Those waivers have had no adverse impacts to the FSS development. To keep focus on quality software deliveries in support of space vehicle testing and operations, the program office will disapprove any waivers which compromise the team's ability to complete the development.

Related GAO Products

Space Acquisitions: Major Space Programs Still at Risk for Cost and Schedule Increases. GAO-08-552T. Washington, D.C.: March 4, 2008.

Space Acquisitions: Space Based Infrared System High Program and Its Alternative. GAO-07-1088R. Washington, D.C.: September 12, 2007.

Defense Acquisitions: Assessments of Selected Weapon Programs. GAO-07-406SP. Washington, D.C.: March 30, 2007.

Space Acquisitions: Actions Needed to Expand and Sustain Use of Best Practices. GAO-07-730T. Washington, D.C.: April 19, 2007.

Space Acquisitions: DOD Needs to Take More Action to Address Unrealistic Initial Cost Estimates of Space Systems. GAO-07-96 (DATE?)

Space Acquisitions: Improvements Needed in Space Systems Acquisitions and Keys to Achieving Them. GAO-06-626T. Washington, D.C.: April 6, 2006.

Space Acquisitions: Stronger Development Practices and Investment Planning Needed to Address Continuing Problems. GAO-05-891T. Washington, D.C.: July 12, 2005.

Defense Acquisitions: Stronger Management Practices Are Needed to Improve DOD's Software-Intensive Weapon Acquisitions. GAO-04-393. Washington, D.C.: March 1, 2004.

Defense Acquisitions: Risks Posed by DOD's New Space Systems Acquisition Policy. GAO-04-0379R. Washington, D.C.: January 29, 2004.

Defense Acquisitions: Improvements Needed in Space Systems Acquisition Policy to Optimize Growing Investment in Space. GAO-04- 253T. Washington, D.C.: November 18, 2003.

Best Practices: Better Matching of Needs and Resources Will Lead to Better Weapon System Outcomes. GAO-01-288. Washington, D.C.: March 8, 2001.

References

[1] GAO, Defense Acquisitions: Space Based Infrared System High Program and its Alternative, GAO-07-1088R (Washington, D.C.: Sept. 12, 2007).

[2] CMMI® (Capability Maturity Model® Integration) is a collection of best practices that helps organizations improve their processes. It was initially developed by product teams from industry, government, and the Software Engineering Institute for process improvement in the development of products and services covering the entire product life cycle from conceptualization through maintenance and disposal. Following the success of CMMI models for development organizations, a CMMI model that addresses the acquisition environment was developed; and can be found within Guidelines for Successful Acquisition and Management of Software-Intensive Systems: Weapon Systems Command and Control Systems Management Information Systems, Department of the Air Force, Software Technology Support Center, (Condensed version) (February 2003).

[3] 10 U.S.C. § 2433, commonly known as "Nunn-McCurdy," generally requires DOD to review programs and report to Congress whenever certain unit cost growth thresholds are reached.

[4] Satellites primarily consist of the payload and the bus. Currently, DOD's buses are custom- made for each space program.

[5] The bus is the platform that provides the power, attitude, temperature control, and other support to the satellite in space.

[6] FSS v1.5 Block 2 Systems Engineering & Incremental Design Review, Lockheed Martin Space Systems Company, Sunnyvale, California.

[7] FFRDCs are unique independent nonprofit entities sponsored and funded by the. government to meet specific long-term technical needs that cannot be met by existing in house or contractor resources. The Aerospace Corporation's FFRDC is sponsored by the Air Force, and provides objective technical analyses and assessments for space programs that serve the national interest. As the FFRDC for nation-security space, Aerospace supports long-term planning and the immediate needs of our nation's military and reconnaissance space programs.

[8] The Software Engineering Institute is a FFRDC that works closely with defense and government organizations, industry, and academia to continuously improve software intensive systems. The Institute's core purpose is to help organizations to improve their software engineering capabilities and to develop or acquire the right software, defect free, within budget and on time.

In: Military Satellites: Issues, Goals and Challenges
Editor: Abel Chirila, pp. 97-110
ISBN: 978-1-60741-238-0

Chapter 5

SPACE ACQUISITIONS: DOD IS MAKING PROGRESS TO RAPIDLY DELIVER LOW COST SPACE CAPABILITIES, BUT CHALLENGES REMAIN*

United States Government Accountability Office

WHAT GAO FOUND

Since GAO last reported on DOD's ORS efforts in 2006, the department has taken several steps toward establishing a program management structure for ORS and executing research and development efforts. On the programmatic side, DOD provided Congress with a plan that lays out an organizational structure and defines the responsibilities of the newly created Joint ORS Office, and describes an approach for satisfying warfighters' needs. DOD has also begun staffing the office. On the research and development side, DOD has launched one of its TacSat satellites—small experimental satellites intended to quickly provide a capability that meets an identified need within available resources—and has begun developing several others. It has also made progress in developing interface standards for satellite buses—the platform that provides power, altitude, temperature control, and other support to the satellite in space—and continued its sponsorship of efforts aimed at acquiring low cost launch vehicles. Despite this progress, it is too early to determine the overall success of these efforts because most are still in their initial phases.

Achieving success in ORS will be challenging. With relatively modest resources, the Joint ORS Office must quickly respond to the warfighter's urgent needs, while continuing research and development efforts that are necessary to help reduce the cost and time of future space acquisitions. As it negotiates these priorities, the office will need to coordinate its efforts with a broad array of programs and agencies in the science and technology, acquisition, and operational communities. Historically, it has been difficult to transition programs from the science and technology environment to the acquisition and operational environment. At this time, DOD lacks a plan that lays out how it will direct its investments to meet current operational needs while pursuing innovative approaches and new technologies.

---

* Excerpted from GAO Report GAO-08-516, dated April 2008.

Why GAO Did this Study

The Department of Defense (DOD) invests heavily in space assets to provide the warfighter with intelligence, navigation, and other information critical to conducting military operations. In fiscal year 2008 alone, DOD expects to spend over $22 billion dollars on space systems. Despite this investment, senior military commanders have reported shortfalls in tactical space capabilities in each recent major conflict over the past decade.

To provide short-term tactical capabilities as well as identify and implement long-term solutions to developing low cost satellites, DOD initiated operationally responsive space (ORS). Following a 2006 GAO review of ORS, the Congress directed DOD to submit a report that sets forth a plan for providing quick acquisition of low cost space capabilities.

This report focuses on the status of DOD's progress in responding to the Congress and is based on GAO's review and analyses of ORS documentation and interviews with DOD and industry officials.

What GAO Recommends

GAO recommends that the Secretary of the Air Force develop an investment plan—approved by stakeholders—that identifies how to achieve future capabilities, establishes funding priorities, and implements mechanisms to measure progress. DOD concurred with the recommendation.

Abbreviations

DARPA	Defense Advanced Research Project Agency
DOD	Department of Defense
GAO	Government Accountability Office
ORS	operationally responsive space
SpaceX	Space Explorations Technologies

April 25, 2008
The Honorable Bill Nelson
Chairman

The Honorable Jeff Sessions
Ranking Member
Subcommittee on Strategic Forces Committee on Armed Services
United States Senate

The Department of Defense (DOD) is investing heavily in large, complex satellites and other space assets to provide the warfighter with communications, intelligence, navigation, missile-warning, and other information critical to conducting military operations. In fiscal year 2008 alone,

DOD expects to spend over $22 billion dollars to develop and procure satellites and other space systems. Yet, for the past two decades, major satellite programs have been beset with significant cost overruns and schedule delays. Moreover, in each major conflict over the past decade, senior military commanders reported shortfalls in tactical space capabilities, such as those intended to provide communications and imagery data to the warfighter in theater.

To help address these issues, DOD recently initiated an effort known as operationally responsive space (ORS). The ORS initiative encompasses several separate endeavors with a goal to provide short-term tactical capabilities as well as identifying and implementing long-term technology and design solutions to reduce the cost and time of developing and delivering simpler satellites in greater numbers. More specifically, these include developing and launching small-size satellites, acquiring lower cost launch systems, developing standardized satellite components, as well as exploring a variety of new common design techniques. Though DOD has tried to make space acquisition more responsive in the past, the current ORS initiative is a concerted effort to create an environment where new concepts and ideas can be fostered and transitioned to users.

We reviewed aspects of the ORS initiative in 2006 and concluded that DOD needed a departmentwide strategy for pursuing low cost, responsive tactical capabilities—both satellite and launch—for the warfighter, and to identify corresponding funding.[1] Subsequently, the Congress directed DOD to submit a report that sets forth a plan for providing quick acquisition of low cost space capabilities.[2] Given the potential these efforts offer for changing the way DOD acquires and fields space capabilities, you asked us to report on the status of DOD's progress to date in implementing the program and assessing associated challenges. We also reviewed DOD's development of a higher level strategy to guide the ORS initiative, an action that we reported separately to you in March 2008. Further, we are reviewing ORS' requirements setting process and its integration into warfighting concepts of operation, which we will report on later in June 2008.[3]

To conduct our work for this report, we reviewed and analyzed ORS documentation and interviewed officials representing the ORS initiative as well as officials from the warfighting, acquisition, science, and technology communities and industry. We conducted this performance audit from May 2007 to February 2008 in accordance with generally accepted government auditing standards. Those standards require that we plan and perform the audit to obtain sufficient, appropriate evidence to provide a reasonable basis for our findings and conclusions based on our audit objectives. We believe that the evidence obtained provides a reasonable basis for our findings and conclusions based on our audit objectives.

RESULTS IN BRIEF

DOD is making progress in putting a program management structure in place for ORS as well as executing ORS-related research and development efforts. On the programmatic side, DOD provided Congress with a plan that lays out an organizational structure and defines the responsibilities of the newly created Joint ORS Office, and describes an approach for satisfying warfighters' needs. DOD has also begun staffing the office. On the research and development side, DOD has launched one of its small- sized satellites and begun developing several others, made progress in developing interface standards for satellite buses, and continued its sponsorship of efforts aimed at acquiring low cost launch vehicles. It is too early to determine how successful these efforts will be because most are still in their initial stages.

As DOD moves forward, it will be challenged on many fronts as ORS is expected to execute a range of efforts within relatively modest resources and amid competing demands and pressure to produce quick results. For example, rapid development and deployment of small satellites is being currently considered as a potential solution for various capability gaps that may occur because of schedule delays on larger acquisition programs. But concentrating efforts to fill just one gap could be relatively expensive and time consuming for the ORS initiative and divert resources from other ORS efforts. For example, some officials we spoke with asserted that not enough attention was being paid to acquiring low cost launch vehicles—a linchpin to reducing satellite development costs in the future. At this time, DOD currently lacks tools needed to negotiate these challenges— primarily, a plan that lays out how it will direct its investments to meet existing gaps while at the same time pursuing innovative approaches and new technologies to rapidly respond to future warfighter needs. To help DOD successfully negotiate this challenge, we are recommending that DOD develop an investment plan.

Background

In May 2003, the Office of Force Transformation began funding small experimental satellites to enhance the responsiveness to the warfighter and to create a new business model for developing and employing space systems.[4] As we have reported over the past two decades, DOD's space portfolio has been dominated by larger space system acquisitions, which have taken longer, cost more, and delivered fewer quantities and capabilities than planned.[5] The ORS initiative is a considerable departure from DOD's large space acquisition approach. The initiative aims to quickly deliver low cost, short-term tactical capabilities to address unmet needs of the warfighter. Unlike traditional large satellite programs, the ORS initiative is intended to address only a small number of unmet tactical needs—one or two—with each delivery of capabilities. It is not designed to replace current satellite capabilities or major space programs in development. Also, the initiative potentially aims to identify and facilitate ways to reduce the time and cost for all future space development efforts.

As we have previously reported, managing requirements so that their development is matched with resources offers an opportunity to mature technologies in the science and technology environment—a best acquisition practice.[6] We also have reported that the ORS initiative could provide opportunities for small companies—who often have a high potential to introduce novel solutions and innovations into space acquisitions—to compete for DOD contracts. Consolidations within the defense industrial base for space programs have made it difficult for such companies to compete. ORS could broaden the defense industrial base and thereby promote competition and innovation.

DOD HAS MADE PROGRESS TO DEVELOP A MANAGEMENT STRUCTURE AND BUILD A TECHNOLOGICAL FOUNDATION FOR ORS

Since we last reported on DOD's ORS efforts in 2006, the department has taken several steps toward establishing a program management structure for ORS and executing research and development efforts. Despite this progress, it is too early to determine the overall success of these efforts because most are still in their initial phases.

Program Management Structure

Congress directed that DOD submit a report that sets forth a plan for the quick acquisition of low cost space capabilities and establish a Joint ORS Office to coordinate and manage the ORS initiative. In the first half of 2007, DOD delivered an ORS plan to Congress and established a Joint ORS Office. DOD created the Joint ORS Office to coordinate and manage specific science and technology efforts to fulfill joint military operational requirements for on-demand space support and reconstitution. In addition, DOD is working with other government agencies to staff the office, developing an implementation plan, and establishing a process for determining which existing requirements for short-term tactical capabilities the office should pursue. Responsiveness is an attribute desired by the entire space community, including the National Aeronautics and Space Administration and the military service laboratories. Most of the efforts under the ORS initiative are being executed by science and technology organizations and other DOD agencies. The office will be responsible for coordinating, planning, acquiring, and transitioning those efforts. Its work is to be guided by an executive committee, comprised of senior officials from DOD, the military services, the intelligence community, and other government agencies. Most requirements for needed short-term tactical capabilities are expected to come through the U.S. Strategic Command. To respond to unmet warfighter needs, ORS requirements will be based on existing validated requirements. Table 1 summarizes the status of some of DOD's efforts related to the management structure.

Table 1. Status of Efforts to Develop a Management Structure by Action

Status	Description
Deliver a plan to Congress as required by the 2007 National Defense Authorization Act	
Completed; DOD submitted the plan in April 2007	• The plan identifies a general approach for establishing the responsive space initiative and establishes an office to coordinate and execute ORS activities. • The Joint ORS Office, located at Kirtland Air Force Base in Albuquerque, New Mexico, was officially activated in May 2007.
Staff the Joint ORS Office	
In progress; 8 of the 20 expected government positions have been filled as of February 2008.	• The office is to be staffed with up to 20 government (military and civilian) positions to be provided by the various military services and agencies associated with responsive space activities. • As of February 1, 2008, 8 government positions were staffed by personnel from the United States Air Force, Army, and Navy; the National Aeronautics and Space Administration; the National Reconnaissance Office; the National Security Agency; the Air National Guard; and Sandia Laboratory. • Additional support is being provided by 14 Federally Funded Research and Development Centers[a] and Systems Engineering and Technical Assistance[b] contractors.

Table 1. (Continued).

Status	Description
Develop an implementation plan	
In progress; the plan is expected to be approved in February 2008	• In July 2007, the Deputy Secretary of Defense directed the Executive Agent for Space to develop an implementation plan that describes how the office will deliver existing space capabilities, evolve new capabilities, and leverage innovative approaches to meet requirements identified by the U.S. Strategic Command. • DOD established a working group with five subgroups tasked with making recommendations on requirements, acquisitions, manpower, lead service, and executive committee. One of the subgroups is tasked with developing a charter for the Executive Committee that would define the roles and responsibilities of the committee members from the various and diverse agencies.
Establish a requirements process	
In progress	• ORS requirements are to be based on existing but unsatisfied validated requirements to respond to combatant commanders' needs. The Joint ORS Office received a request from U.S. Strategic Command to develop a communications capability that cannot be satisfied by the current ultra high frequency system. The requested capability falls within the validated requirements set for the Mobile User Objective System, currently under development[c]

Source: DOD data and GAO analysis.

[a] Federally Funded Research and Development Centers conduct research for the U.S. government.

[b] Systems Engineering and Technical Assistance contractors are civilian employees of government contractors who are contracted to assist DOD components and acquisition programs.

[c] The Navy's Mobile User Objective System is expected to provide low data rate voice and data communications capable of penetrating most weather, foliage, and man-made structures beginning in late 2009.

Research and Development Efforts

DOD is continuing to make progress in developing TacSats—its small experimental satellite projects. In addition, DOD is funding research efforts by industry to facilitate the development of future capabilities and is working with industry and academia to develop standards for building satellite components. Finally, DOD is working to improve the capabilities of existing small launch vehicles and providing some funding for future launch vehicles.

Development of Small-Sized Satellites

The TacSat experiments aim to quickly provide the warfighter with a capability that meets an identified need within available resources—time, funding, and technology. Limiting the TacSats' scope allows DOD to trade off higher reliability and performance[7] for speed, responsiveness, convenience, and customization. Once each TacSat satellite is launched, DOD plans to test its level of utility to the warfighter in theater. If military utility is established, DOD will assess the acquisition plan required to procure and launch numerous TacSats—forming constellations—to provide wider coverage over a specific theater. As a result, each satellite's capability does not need to be as complex as that of DOD's larger satellites and does not carry with it the heightened consequence of failure as if each satellite alone were providing total coverage. DOD currently has five TacSat experiments in different stages of development (see table 2).

Table 2. Status of Efforts to Develop and Demonstrate TacSat Experimental Satellites

Status	Description
TacSat 1	
In progress; satellite has been developed but not demonstrated.	• The Naval Research Laboratory led a year-long effort to develop TacSat 1, at a cost of $23 million. • TacSat 1 was completed in May 2004, but has yet to be demonstrated because of delays incurred with the development and testing of a low cost launch vehicle. • Given the launch delay, the Naval Research Laboratory decided to add a new sensor—an automated identification system to support maritime missions. The new sensor and other new capabilities are estimated to cost $10.5 million. • The Joint ORS Office, the Navy, Coast Guard, and Department of Defense Research and Engineering are currently working to develop a cost sharing agreement.
TacSat 2	
Complete; satellite developed and demonstrated in 2006 through 2007.	• TacSat 2 development, led by the Air Force Research Laboratory, was completed in 29 months at a cost of $39 million. • Its payload includes tactical imaging and radio frequency equipment, and an automated identification sensor. • TacSat 2 was launched in December 2006 on a Minotaur I launch vehicle and participated in military exercises during the summer of 2007. • The Air Force Research Laboratory ended the demonstration in December 2007.
TacSat 3	
In progress; satellite expected to be launched in August 2008.	• The Air Force Research Laboratory is leading the effort to develop TacSat 3 which will provide the first implementation of selected bus standards. • Its primary payload is a hyperspectral imager for tactical targeting of camouflaged and hard-to-detect targets. • The cost is estimated to be $62.7 million, and the planned launch date is August 2008 on a Minotaur I launch vehicle.
TacSat 4	
In progress; expected to be launched in September 2009.	• The Naval Research Laboratory is leading the effort to develop TacSat 4, which will include equipment to demonstrate mobile data communications services, friendly forces tracking, and data relay from terrestrial sensor. • It will also evaluate the DOD system bus standards effort in a realistic launch-andflight operations environment. • The cost is estimated to be $114 million, and the planned launch date is September 2009 on a Minotaur IV launch vehicle.
TacSat 5	
In progress; launch date to be determined.	• The Army Space and Missile Defense Center, the Joint ORS Office, the Air Force Research Laboratory, and Space and Missile Systems Center will lead the development of TacSat 5. • Payload experiments have not been finalized. • The cost and schedule are to be determined.

Source: DOD data.

In addition, DOD is sponsoring the development of new capabilities provided mostly by the small satellite industry. These efforts include the ORS Payload Technology Initiative, which awarded 15 contracts to satellite industry contractors for payload technology concepts that may be developed in the future. The Air Force has been funding additional research conducted by small technology companies that could provide ORS capabilities, such as faster ways of designing satellites, and identifying the types and characteristics of components based on mission requirements.

Efforts to Develop and Test Satellite Interface and Bus Standards

DOD is also working to establish standards for the "bus"—the platform that provides power, attitude, temperature control, and other support to the satellite in space.[8] Establishing interface standards for bus development would allow DOD to create a "plug and play" approach to building satellites—similar to the way personal computers are built. According to DOD officials, interface standards would allow the development of modular or common components and would facilitate building satellites—both small and large—more quickly and at a lower cost. DOD's service laboratories, industry, and academia have made significant progress to develop satellite bus standards. The service labs expect to test some standardized components on the TacSat 3 bus and system standards on the TacSat 4 bus. Table 3 provides a description of the bus standardization efforts and their status.

Table 3. Status of Efforts to Develop and Test Satellite Interface and Bus Standards

Status	Description
Demonstrate "plug and play" architecture	
In progress.	• The Air Force Research Laboratory is demonstrating capabilities of a "plug and play" architecture that accelerates satellite production through the use of standardized computer-like ports to connect standardized components. • For example, the Air Force Research Laboratory is developing standard bus panels with an embedded wiring system and standard ports for easy assembly and integration.
Determine class of ORS satellites needed	
Complete; report issued in March 2005.	• This effort, led by the Massachusetts Institute of Technology Lincoln Laboratory, determined the class of satellite needed to be militarily useful.
Develop interface standards, and develop TacSat 3 bus	
In progress.	• This effort, being led by the Air Force Research Laboratory, is to provide interface standards between the bus and the payload, and to rapidly provide a bus for the TacSat 3 hyperspectral payload experiment.
Develop bus standards for initial blocks of operational satellites, and transition plan	
In progress.	• This effort, being led by the Naval Research Laboratory and the John Hopkins University Applied Physics Laboratory, is using a broad consortium of industry and government members to develop bus standards along with the associated costs and business considerations for acquisition. • The consortium is also working to develop a plan that explains how to transition from a science and technology environment to an acquisition phase. The transition plan is in final coordination within DOD.
Acquire initial standardized satellite buses	
Slated to begin after bus standards have been developed.	• This effort, to be carried out by the Joint ORS Office, will be the acquisition of the initial block of standard satellite buses.

Source: DOD data.

Launch Infrastructure

To get new tactical space capabilities to the warfighter sooner, DOD must secure a small, low cost launch vehicle on demand. Current alternatives include Minotaur launch vehicles, ranging in cost from about $21 million to $28 million, and an Evolved Expendable Launch Vehicle—DOD's primary satellite launch vehicles—at an average cost of roughly $65 million (for medium and intermediate launchers). DOD is looking to small launch vehicles, unlike current systems, that could be launched in days, if not hours, and whose cost would better match the small budgets of experiments. Both DOD and private industry are working to develop small, low cost, on-demand launch vehicles. Notably, DOD expects the Defense Advanced Research Projects Agency's (DARPA) FALCON launch program to flight-test hypersonic technologies and be capable of launching small satellites such as TacSats. In addition to securing low cost launch vehicles, DOD plans to acquire a more responsive, reliable, and affordable launch tracking system to complement the existing launch infrastructure. Table 4 describes DOD's efforts to develop a launch infrastructure and their status.

Table 4. Efforts to Develop a Launch Infrastructure

Status	Description
Develop low cost launch vehicles through DOD's FALCON program	
In progress.	• DARPA, along with the Air Force, established FALCON, to accelerate efforts to develop a vehicle that can send 1,000 pounds to low-earth orbit for less than $5 million with an operational cost basis of 20 flights per year for 10 years. • DARPA contracted with Space Exploration Technologies (SpaceX) to develop a twostage vehicle for launches into space. According to DOD, SpaceX completed its activity for the FALCON program with a test launch of their rocket in March 2007. • For fiscal year 2008, the Appropriations conferees encouraged ORS to continue to work with the FALCON program to develop a small launch vehicle. Another SpaceX flight demonstration is expected to occur in mid 2008. • According to DARPA, between fiscal years 2003 and 2007, it provided SpaceX with a total of more than $15.6 million dollars for advanced technology development, facilities, test-range and mission support, and program office support.
Develop new launch vehicle capabilities through the Air Force Research Laboratory	
In progress.	• The Air Force Research Laboratory continues its work on spacecraft payload shock protection and noise reduction technologies during launch. • The laboratory has developed technologies to reduce vibration in launch vehicles that have been used on the Minotaurs. • The laboratory has also funded science and technology research on solid rocket motors.
Develop other elements that support responsive launch systems	
In progress	• In September 2006, the DOD Executive Agent for Space signed a memorandum stating that by January 1, 2011, all DOD, civil, and commercial vehicles launched will need to be tracked through use of the Global Positioning System.a The DOD Executive Agent for Space sees this as the first logical step towards a space-based control system that is more responsive, reliable, and affordable than the current terrestrially based system.

Source: DOD data.

[a] The Global Positioning System is a space-based radio-positioning system nominally consisting of a 24-satellite constellation providing navigation and timing data to military and civilian users worldwide.

ORS Challenges Magnified by Lack of a Plan to Guide Investments and Efforts

DOD faces several challenges in succeeding in its ORS efforts. With relatively modest resources, the Joint ORS Office must quickly respond to the warfighter's urgent needs, including gaps in capabilities, as well as continue its longer-term research and development efforts that are necessary to help reduce the cost and time of future space acquisitions. As the office negotiates these priorities, it will need to coordinate its efforts with a broad array of programs and agencies in the science and technology, acquisition, and operational communities. Historically it has been difficult to transition programs initiated in the science and technology environment to the acquisition and operational environment. At this time, DOD lacks tools which would help the program office navigate within this environment—primarily, a plan that lays out how the office will direct its investments to meet current operational needs while at the same time pursuing innovative approaches and new technologies.

The Joint ORS Office has a budget totaling about $646 million for fiscal years 2008 through 2013 and with no more than 20 government staff. These resources are relatively modest when compared with the resources provided major space programs. For example, the ORS fiscal year 2008 budget represents less than 12 percent of the budget of the Transformational Satellite Communications System program[9] which is in the concept development phase, and staffing is about a quarter of that program's staff. While the Joint ORS Office's responsibilities are not the same as those of large, complex acquisition programs, it is expected to address urgent tactical needs that have not been met by the larger space programs. At this time, for example, the office has been asked to develop a solution to meet current communications shortfalls that cannot be met by the current Ultra High Frequency Follow-On satellite system.[10] And, while the office has not yet been asked, officials have told us that ORS could potentially satisfy a gap in early missile warning capabilities because of delays in the Space Based Infrared Systems program, as well as gaps in communications and navigation capabilities. Taking on any one of these efforts will be challenging for ORS as there are constraints in available technologies, time, money, and other resources that can be used to fill capability gaps.

At the same time, the Joint ORS Office will be pressured to continue to sponsor longer term research and development efforts. According to the Air Force Research Laboratory, the average cost of a small satellite is about $87 million. This is substantially higher than the target acquisition cost of about $40 million for an integrated ORS satellite in the 2007 National Defense Authorization Act.[11] In addition, concerns are being expressed that not enough funding and support are being devoted to acquiring low cost launch capabilities. Some government and industry officials believe that achieving such capabilities is a linchpin to reducing satellite development costs in the future. The current alternatives for launching ORS satellites—an Evolved Expendable Launch Vehicle and Minotaur launch vehicles—do not meet DOD's low cost goal. DARPA expects its responsive launch capabilities, currently in development, will total about $5 million to produce—a significantly lower cost than that of current capabilities. However, in order to achieve the lower cost launch capability, DOD will have to continue to fund research beyond the $15.6 million already spent on advanced technology development, facilities, test- range and mission support, and program office support.

To execute both its short- and long-term efforts, the Joint ORS Office will also need to gain cooperation and consensus from a diverse array of officials and organizations. These include science and technology organizations, the acquisition community, the U.S. Strategic Command, the intelligence community, and industry. We have previously reported on difficulties DOD has encountered in bringing these organizations together, particularly when it comes to setting requirements for new acquisitions and transitioning technologies from the science and technology community to acquisition programs. As a new and relatively small organization, the Joint ORS Office may well find it does not have the clout to gain cooperation and consensus on what short- and long-term projects should get the highest priority.

Despite the significant expectations placed on the Joint ORS Office and the challenges it faces, DOD does not have an investment plan to guide its ORS decisions. DOD has begun to develop an ORS strategy that is to identify the investments needed to achieve future capabilities. However, the strategy is not intended to become a formalized investment plan that would (1) help DOD identify how to achieve these capabilities, (2) prioritize funding, and (3) identify and implement mechanisms to enforce the plan. At the same time, there are other science and technology projects in DOD's overall space portfolio competing for the same resources, including those focused on discovering and developing technologies and materials that could enhance U.S. superiority in space. Further, as DOD's major space acquisition programs continue to experience cost growth and schedule delays, DOD could be pressured to divert funds from ORS. We have previously recommended that DOD prioritize investments for both its acquisitions and science and technology projects—the ORS plan could be seamlessly woven into an overall DOD investment plan for space. However, DOD has yet to develop this overall investment plan.

Conclusions

Providing the warfighter with needed space capabilities in a fiscally constrained and rapidly changing technological environment is a daunting task. ORS provides DOD with a unique opportunity to work outside the typical acquisition channels to more quickly and less expensively deliver these capabilities. However, even at lower costs, pressure on ORS funding will come in DOD's competition for its resources. As DOD moves forward, decisions on using constrained resources to meet competing demand will need to be made and reevaluated on a continuing basis. Until DOD develops an investment plan, it will risk forgoing an opportunity to get continuing success out of the ORS initiative.

Recommendation for Executive Action

To better ensure that DOD meets the ORS initiative's goal, we recommend that the Secretary of the Air Force develop an investment plan to guide the Joint ORS Office as it works to meet urgent needs and develops a technological foundation to meet future needs. The plan should be approved by the stakeholders and identify how to achieve future capabilities, establish funding priorities, and identify and implement mechanisms to ensure progress is being achieved.

AGENCY COMMENTS

We provided a draft of this report to DOD for review and comment. DOD concurred with our findings and our recommendation but clarified that the Secretary of the Air Force, specifically the Executive Agent for Space, would be responsible for developing an investment plan since the Under Secretary of the Air Force position is vacant. Full comments can be found in appendix I.

SCOPE AND METHODOLOGY

To assess DOD's progress to date in implementing its ORS goal and addressing associated challenges, we interviewed and reviewed documents from officials in Washington, D.C., at the Office of the Deputy Under Secretary of Defense for Advanced Systems and Concepts; National Security Space Office; Office of the Director of Defense Research and Engineering; Office of the Director of Program Analysis and Evaluation; Office of the Joint Chiefs of Staff; the U.S. Naval Research Laboratory; and the Office of the Assistant Secretary of the Navy for Research, Development and Acquisition. We also interviewed and reviewed documents from officials in Virginia at the Office of the Assistant Secretary of Defense for Networks Information and Integration; Office of the Under Secretary of the Air Force; Defense Advanced Research Project Agency; and U.S. Army Space and Missile Defense Command. In addition, we interviewed and reviewed documents from officials at the Navy Blossom Point Satellite Tracking Facility in Maryland; Air Force Space Command, Peterson Air Force Base, Colorado; Space and Missile Systems Center, Los Angeles Air Force Base, California; the U.S. Strategic Command, Offutt Air Force Base, Nebraska; and the Air Force Research Laboratory and Joint Operationally Responsive Space Office, Kirtland Air Force Base, New Mexico. We also interviewed officials from the National Aeronautics and Space Administration, Washington, D.C., and industry representatives involved in developing small satellites and commercial launch vehicles. We reviewed and analyzed the documents that we received.

We will send copies of the letter to the Department of Defense and other interested congressional committees. We will also make copies available to others upon request. In addition, the report will be available at no charge on the GAO Web site at http://www.gao.gov.

Should you or your staff have any questions on matters discussed in this report, please contact me at (202) 512-4859 or chaplainc@gao.gov. Contact points for our Offices of Congressional Relations and Public Affairs may be found on the last page of this report. Principal contributors to this report were Art Gallegos, Assistant Director; Maria Durant; Jean Harker; Arturo Holguin; and Karen Sloan.

Cristina Chaplain
Director
Acquisition and Sourcing Management

APPENDIX I: COMMENT FROM THE DEPARTMENT OF DEFENSE

OFFICE OF THE ASSISTANT SECRETARY OF DEFENSE
6000 DEFENSE PENTAGON
WASHINGTON, DC 20301-6000

NETWORKS AND
INFORMATION INTEGRATION

Ms. Christina T. Chaplain
Director, Acquisition and Sourcing Management
Government Accountability Office
441 G Street, N.W.
Washington, D.C. 20548

Dear Ms. Chaplain:

The Department of Defense (DoD) concurs with the GAO Draft Report, GAO-08-516, 'SPACE ACQUISITIONS: DoD Is Making Progress to Rapidly Deliver Low Cost Space Capabilities, but Challenges Remain,' dated March 14, 2008 (GAO Code 120648). For clarification, the investment plan should be worked by the DoD Executive Agent for Space, currently the Secretary of the Air Force, and not the Under Secretary of the Air Force as that position is vacant. Thank you for the opportunity to review and comment on this report.

Sincerely,

Steven M. Huybrechts
Director, Space Programs and Policy

REFERENCES

[1] GAO, Space Acquisitions: DOD Needs a Departmentwide Strategy for Pursuing Low- Cost, Responsive Tactical Space Capabilities, GAO-06-449 (Washington, D.C.: Mar. 14, 2006).

[2] The John Warner National Defense Authorization Act for Fiscal Year 2007, Pub. L. No. 109-364, § 913(c) (2006).

[3] GAO, Defense Space Activities: National Security Space Strategy Needed to Guide Future DOD Space Efforts, GAO-08-431R (Washington, D.C.: Mar. 27, 2008).

[4] Department of Defense, Quadrennial Defense Review Report (September 2001).

[5] GAO, Space Acquisitions: Improvements Needed in Space Systems Acquisitions and Keys to Achieving Them, GAO-06-626T (Washington, D.C.: Apr. 6, 2006).

[6] GAO, Best Practices: Stronger Practices Needed to Improve DOD Technology Transition Processes, GAO-06-883 (Washington, D.C.: Sept. 14, 2006).

[7] Existing systems have been designed for longer life and increased reliability, requiring years to develop and a significant investment of resources.

[8] Satellites primarily consist of the payload and the bus. Currently, DOD's buses are custom-made for each space program.

[9] The Air Force's Transformational Satellite Communications System is expected to provide high data rate military satellite communications services to DOD users.

[10] This system is expected to be replaced by the Navy's Mobile User Objective System.

[11] The John Warner National Defense Authorization Act for Fiscal Year 2007, Pub. L.No. 109-364 § 913(b)(1)(2006)(codified at 10 U.S.C. § 2273a(e)(5)).

INDEX

A

absorption, 11
academic, 1, 15
access, 2, 12, 13, 21
accountability, viii, 77, 80, 83, 86, 87
acid, 11
acquisition phase, 104
acquisitions, 22, 79, 97, 100, 102, 106, 107
aerospace, 12
Afghanistan, 2
agriculture, 2
air, vii, 15, 19
AL, 16
allies, vii, 13, 53, 54, 81
alternative, viii, 54, 55, 57, 58
alternatives, viii, 77, 80, 83, 92, 105, 106
aluminum, 13
appendix, 90, 91, 108
application, 4, 11, 13, 14
Army, 55, 101, 103, 108
assessment, 3, 56, 57, 81, 86, 91
assets, vii, 2, 3, 13, 22, 98
assumptions, 57, 80, 90, 91
asynchronous, 84
atmosphere, 2, 11
atoms, 8, 9, 10
attacks, 14
auditing, viii, 23, 54, 79, 92, 99
authority, 86, 87
availability, 57
awareness, 76

B

bacterium, 8
basic research, 10, 14
batteries, 6, 21
beams, 4
Beijing, 16
benefits, 3, 11, 14
benzene, 9
Best Practice, v, 19, 95, 109
biomaterials, 8
blindness, 4
blocks, 21, 104
bottom-up, 8
brain, 10
broad spectrum, 80, 89
building blocks, 7, 9
burn, 11
buses, viii, 96, 97, 99, 104, 110
business model, 100

C

campaigns, 6
Capacity, 17
carbon, 9, 10, 11, 12, 13, 17, 18
carbon atoms, 9, 10
carbon nanotubes, 9, 10, 11, 12, 13
catalyst, 11
cell, 10
cell membranes, 10
channels, 107
Chief of Staff, 2
China, 2, 4, 5, 13, 16
chiral, 9
civilian, 101, 102, 105
clusters, 9
CMOS, 9, 12
Co, 17
Coast Guard, 103
coatings, 3, 10, 12, 14

coding, 77, 80, 83, 84, 85, 88, 89
collaboration, 21
Colorado, 23, 108
commerce, 2
commercialization, 12
Committee on Armed Services, 19, 53, 92, 98
communication, 21, 87
communications satellites, vii, 2
communities, 97, 99, 106
community, 2, 5, 8, 22, 101, 107
competition, 100, 107
complement, 105
complexity, 22, 55, 84, 87, 88, 90, 91
components, 4, 6, 7, 8, 12, 55, 57, 81, 99, 102, 104
composition, 10
conceptualization, 95
conduction, 9
conductive, 7, 12
conductivity, 3, 10, 11, 12
confidence, 21, 77, 78, 80, 87, 88, 90, 91
conflict, vii, 2, 98, 99
Congress, iv, viii, 16, 18, 80, 90, 96, 97, 98, 99, 101
consensus, 85, 106
constant rate, 5
constraints, 22, 106
construction, 11
contingency, 5
contractors, 22, 80, 87, 101, 102, 103
contracts, 55, 57, 100, 103
control, vii, viii, 2, 3, 4, 6, 8, 12, 21, 55, 56, 77, 79, 80, 81, 82, 83, 86, 90, 92, 96, 97, 104, 105
copper, 12
costs, 11, 21, 55, 56, 100, 104, 106, 107
covalent, 10
covalent bond, 10
coverage, 102
covering, 95
CRC, 17
credit, 22, 90
cryogenic, 76

D

danger, 5
DARPA, 12, 98, 105, 106
data communication, 102, 103
data processing, 55, 81
database, 80, 85, 89
decisions, 20, 22, 23, 107
defects, 11
defense, viii, 54, 56, 57, 58, 96, 100
Defense Advanced Research Projects Agency (DARPA), 12, 18, 105
deficiency, 84
definition, 8, 12, 55
degradation, 11
delivery, 21, 56, 57, 82, 83, 84, 87, 89, 100
demand, 101, 105, 107
denial, 2, 4
density, 9, 12
Department of Defense (DOD), vii, viii, 1, 16, 19, 23, 24, 53, 58, 59, 78, 79, 92, 93, 98, 103, 108, 109
Desert Storm, 2
destruction, 3, 4, 6, 14
detection, viii, 54, 76, 78, 79, 82
diamond, 10
direct investment, 90
discipline, 86, 90
discretionary, 22
discretionary spending, 22
disposition, 84
distortions, 9
distributed applications, 84
distribution, viii, 77, 80
dominance, 3, 13
dosage, 12
draft, 23, 58, 90, 108
DuPont, 10
durability, 12
duration, 12

E

early warning, vii, 53
earth, 4, 5, 6, 7, 12, 55, 78, 81, 105
electric field, 12
electrical conductivity, 3, 11, 12
electrical power, 7
electrical properties, 12
electricity, 6, 9
electromagnetic, 4, 11, 12, 13
electron, 9
electrons, 9, 10, 11
email, 16, 17, 18
emitters, 11
employees, 102
energy, vii, 1, 2, 3, 4, 5, 6, 7, 9, 10, 11, 12, 13, 14
enthusiasm, 9
environment, 2, 6, 7, 10, 12, 20, 21, 57, 83, 95, 97, 99, 100, 103, 104, 106, 107
Environmental Protection Agency, 10, 17
EPA, 17
estimating, 92
etching, 8
execution, viii, 55, 77, 80, 87

expenditures, 56
expert, iv
expertise, 92
exposure, 7

F

fabric, 12
fabrication, 11, 12, 56
failure, viii, 5, 77, 78, 79, 80, 87, 90, 102
FAS, 16
fault detection, 82
February, 15, 18, 76, 95, 99, 101, 102
fiber, 13
filters, 4
fish, 10
FL, 17
flexibility, 9
flight, viii, 13, 56, 77, 78, 79, 80, 81, 82, 83, 85, 86, 89, 90, 92, 105
flow, 89
fluctuations, 7
forests, 11
freedom, 2
funding, viii, 13, 22, 23, 54, 55, 58, 77, 80, 82, 90, 98, 99, 100, 102, 103, 106, 107
funds, 10, 22, 56, 78, 80, 91, 107

G

GAO, viii, 19, 20, 53, 55, 76, 77, 78, 82, 84, 85, 86, 88, 91, 95, 97, 98, 102, 108, 109
generation, 55
Germany, 17
Global Positioning System, 6, 22, 105
globalization, 5, 13
goals, viii, 22, 23, 54, 55, 56, 76, 78, 79, 80, 82, 87, 89, 90, 91
government, vii, viii, 1, 2, 4, 10, 14, 15, 23, 54, 89, 91, 95, 96, 99, 101, 102, 104, 106
Government Accountability Office, v, 19, 53, 77, 97, 98
graphene sheet, 9
ground-based, vii, 1, 3, 4, 5, 7, 13, 14, 15
groups, vii, 2, 21
growth, 8, 96, 107
guidance, 87

H

handling, 23
hardness, 3
hazards, 10
health, 2, 10, 79, 82, 83, 92
heat, 6, 7, 10, 11
heavy metal, 13
high risk, 58, 88
high-risk, 85
hospitals, 2
host, 3, 55, 81
House, 18, 19, 53, 92
human, 8, 10
humans, 10
hybrid, 10
hydrogen, 8
hydrogen atoms, 8

I

id, 18
identification, 103
imagery, vii, 2, 6, 99
imaging, 103
implementation, 86, 101, 102, 103
inclusion, 21
industrial, 100
industry, 11, 95, 96, 98, 99, 102, 103, 104, 105, 106, 107, 108
Information System, 95
infrared, vii, 53, 54, 55, 57, 76, 81
infrastructure, 3, 58, 105
injury, iv
innovation, 100
insight, 21, 87
instruments, 6
insurance, 56
integrated circuits, 12
integration, viii, 21, 22, 56, 77, 80, 81, 83, 84, 85, 87, 88, 89, 91, 99, 104
intelligence, vii, viii, 6, 20, 22, 54, 76, 79, 98, 101, 107
Intelligence Community, 21
intelligence gathering, 76
interaction, 86
interface, viii, 77, 80, 84, 97, 99, 104
interference, 3, 13
interview, 18
interviews, 23, 98
investment, vii, 1, 5, 10, 14, 22, 58, 98, 100, 107, 108, 110
Iraq, 2
isomers, 11

J

January, 2, 6, 13, 15, 16, 17, 18, 22, 79, 83, 89, 95, 105
John Warner National Defense Authorization Act, 109, 110
Joint Chiefs, 108

K

killing, 4

L

labor, 88
laser, vii, 4, 5, 6, 20, 21
lasers, 3, 4, 5
lattice, 9
lead, 10, 90, 102, 103
leadership, 87, 90
leakage, 13
LEO, 5, 6
lesions, 10
life cycle, 20
likelihood, 23
Lincoln, 104
links, vii, 20
lithium, 21
Lockheed Martin, 58, 78, 79, 80, 81, 82, 83, 84, 85, 86, 87, 88, 89, 92, 96
long distance, 4
long-term, 9, 14, 22, 96, 98, 99, 106
Los Angeles, 23, 92, 108
low earth orbit, 4
low temperatures, 76
lungs, 10

M

magnetic, iv, 8
maintenance, 95
malicious, 5
management, viii, 22, 23, 54, 56, 58, 77, 79, 80, 82, 83, 84, 85, 86, 87, 90, 91, 92, 97, 99, 101
man-made, 102
manpower, 102
maritime, 103
market, 10, 11
Maryland, 23, 108
Massachusetts, 104
Massachusetts Institute of Technology, 104
material degradation, 11
matrix, 12
maturation, 20, 21
measurement, 9
measures, 2, 90
melting, 4
membranes, 11
memory, 7, 13
metals, 9, 10
Mexico, 101, 108
microelectronics, 8
micrometer, 8
microprocessors, 3, 13
microwave, 4, 5
microwaves, 3, 4, 7
military, vii, 2, 3, 4, 5, 6, 10, 12, 13, 14, 20, 53, 54, 58, 81, 96, 98, 101, 102, 103, 105, 110
mirror, 86
missile launches, vii, 53, 56
missions, viii, 7, 54, 103
ML, 18
modeling, 86
models, 87, 95
modulus, 9
momentum, 10
money, 20, 87, 106
morale, 87
motion, 9
motivation, 8
motors, 105

N

nanoelectronics, 8, 9, 12
nanomaterials, 3, 8, 10, 13, 14
nanometer, 7, 8, 9
nanometer scale, 8
nanometers, 8, 9
nanoparticles, 8, 10
nanoscale materials, 8
nanostructures, 8, 9
nanotechnology, vii, 1, 3, 7, 8, 9, 10, 14
nanotube, 9, 10, 11, 12
nanotubes, 9, 10, 11, 12, 13
nation, vii, 1, 2, 3, 15, 53, 96
national, vii, 20, 96
National Aeronautics and Space Administration, 101, 108
National Defense Authorization Act, 82, 101, 106
National Guard, 101
National Research Council, 16, 17, 18
natural, 5
navigation system, 6

Navy, 101, 103, 108
Nebraska, 108
network, 2, 22
networking, 20
New Mexico, 101, 108
New York, iii, iv
next generation, 8, 56
nickel, 13
Nielsen, 15
noise, 105
normal, 21
nuclear, 2, 5

O

obsolete, 17
Office of Management and Budget, 91
Offices of Congressional Relations and Public Affairs, 23, 58, 108
one dimension, 8
online, 86
operating system, 84
optical, 4, 6, 21
optimism, 23
orbit, 2, 5, 6, 12, 55, 57, 78, 81, 105
organization, 21, 86, 107
organizations, 10, 20, 21, 22, 95, 96, 101, 106
oversight, viii, 55, 77, 79, 80, 83, 86, 87, 92

P

paper, vii, 1, 3, 7, 15
particles, 8, 10, 12, 13
passive, 3, 4, 7
pay off, vii, 1, 14
Pentagon, 5
performance, viii, 12, 21, 23, 54, 55, 56, 57, 58, 76, 79, 81, 86, 91, 92, 99, 102
permit, vii, 2
personal, 104
personal computers, 104
photovoltaic, 6
photovoltaic cells, 6
physical properties, 10
physics, 6, 8
planar, 9
planning, 2, 5, 57, 92, 96, 101
plasma, 11
platforms, 20
play, 3, 22, 104
polymer, 12
polymer materials, 12
polymer matrix, 12
poor, 55, 79, 81, 84, 86, 90, 91
poor performance, 81, 91
poor relationships, 90
population, 10
portfolio, 22, 58, 100, 107
ports, 104
power, viii, 4, 6, 7, 14, 16, 82, 96, 97, 104
powers, 2
predictability, 91
President Clinton, 4
pressure, 11, 88, 100, 107
priorities, 22, 90, 97, 98, 106, 107
private, 105
product life cycle, 95
production, 8, 12, 22, 104
productivity, 80, 84, 87, 88, 92
program, viii, 11, 12, 13, 17, 20, 21, 22, 23, 53, 54, 55, 56, 57, 58, 76, 77, 78, 79, 80, 81, 82, 83, 86, 87, 88, 89, 90, 91, 92, 96, 97, 99, 101, 105, 106, 110
proliferation, 4
promote, 12, 100
propagation, 5
property, iv, 1
protection, 7, 11, 18, 105
public, vii, 2
pulse, 7
pulses, 5
pumps, 7
purification, 11

Q

quantum, 8, 9
quantum well, 8

R

race, 2
radar, 21
radiation, 3, 4, 6, 7, 12, 13, 14, 16, 18
radio, vii, 4, 20, 103, 105
radiofrequency, 4
range, vii, 2, 4, 8, 11, 12, 80, 89, 100, 105, 106
rat, 10
reactivity, 10
reality, 2
reduction, 8, 55, 105
reflection, 11
reflectivity, 11
refractive index, 12

refrigeration, 76
relationships, 86, 87, 90
reliability, 17, 90, 102, 110
Republicans, 18
research, vii, viii, 1, 2, 3, 5, 8, 9, 10, 11, 12, 13, 14, 15, 21, 57, 78, 83, 84, 97, 99, 101, 102, 103, 105, 106
research and development, vii, viii, 1, 3, 5, 10, 12, 14, 78, 83, 84, 97, 99, 101, 102, 106
researchers, 10, 14
reserves, 54, 56, 79, 91
resilience, 9
resolution, 84
resources, viii, 3, 20, 21, 22, 23, 54, 58, 85, 89, 96, 97, 100, 102, 106, 107, 110
responsibilities, viii, 22, 86, 97, 99, 102, 106
responsiveness, 100, 102
restructuring, 54, 56
RF, 5
rings, 9
risk, 21, 23, 55, 56, 57, 58, 77, 78, 80, 84, 85, 86, 87, 88, 89, 90, 92, 107
risk management, 23
risks, 10, 21, 55, 56, 57, 58, 78, 79, 81, 85, 87, 88, 89, 91, 92
road map, 55
robustness, 83, 84, 86
Russia, 4, 5, 13
Russian, 4, 5

S

S&T, 21
safety, 79, 82, 83, 92
satellite, vii, viii, 1, 2, 3, 4, 5, 6, 7, 12, 14, 20, 22, 23, 54, 55, 56, 57, 78, 79, 80, 81, 82, 83, 84, 87, 89, 90, 92, 96, 97, 99, 100, 102, 103, 104, 105, 106, 110
satellite orbits, 3
satellite service, 2
scalable, 12
scaling, 9
scattering, 12
schema, 84
school, 2, 8
scientists, vii, 2, 11
SD, 79
search, 16
Secretary of Defense, 23, 55, 57, 58, 78, 79, 80, 84, 88, 90, 91, 92, 102, 108
secure communication, vii, 19
security, vii, 2, 96
selecting, 92
self-assembly, 9
semiconductor, 12
Senate, 19, 53, 92, 98
sensing, 6
sensors, 4, 6, 55, 56, 76, 81
services, iv, vii, 2, 95, 101, 103, 110
severity, 84
shape, 9
sharing, 5, 13, 22, 103
shock, 105
short-term, 22, 98, 99, 100, 101
Siemens, 11, 17
signs, 5
silica, 8, 12
silicon, 9
simulation, 85
single-wall carbon nanotubes, 9, 10
software, vii, viii, 5, 20, 21, 22, 54, 55, 56, 58, 77, 78, 79, 80, 82, 83, 84, 85, 86, 87, 88, 89, 90, 91, 92, 96
software code, 22, 84
solar, 4, 6, 21
solar cell, 6, 21
solar cells, 6, 21
solar energy, 6
solar panels, 4, 6
solid-state, 8
solutions, viii, 7, 10, 77, 80, 84, 98, 99, 100
Soviet Union, 4, 5
space environment, 7
Space Radar, v, vii, 19, 20, 21, 22, 23, 25, 57
Space Tracking and Surveillance System, 55, 81
spectrum, vii, 2, 12, 21, 80, 89
speed, 4, 5, 6, 102
speed of light, 4
sponsor, 106
SR, 17
stability, 9
staffing, viii, 87, 97, 99, 106
stages, 14, 100, 103
stakeholders, 85, 98, 107
standardization, 104
standards, viii, 23, 54, 58, 79, 92, 97, 99, 102, 103, 104
steel, 9
stiffness, 9
strategies, 23, 56
strength, 9, 12
subgroups, 102
suffering, 9
summer, 103
sunlight, 6
superiority, vii, 2, 3, 13, 19, 107

surface area, 6
surveillance, vii, viii, 20, 54, 55, 79, 81
surviving, 12
susceptibility, 5
switching, 12
symptoms, 7
synchronous, 84
systems, vii, viii, 2, 3, 4, 5, 6, 7, 8, 9, 13, 14, 15, 19, 20, 22, 23, 55, 56, 76, 77, 79, 80, 83, 84, 87, 88, 89, 92, 96, 98, 99, 100, 105, 110

T

tangible, 76
targets, 4, 103
technological change, 7
technology, 2, 4, 5, 7, 8, 12, 13, 20, 21, 23, 55, 56, 57, 58, 76, 90, 97, 99, 100, 101, 102, 103, 104, 105, 106, 107
telecommunications, 2
television, 11
temperature, viii, 7, 96, 97, 104
test procedure, 76
Texas, 16
theory, 11
thermal energy, 6
thermal properties, 11
thinking, 23
threat, vii, 1, 3, 4, 7, 10, 13
threatened, 2, 13
threatening, 4
threats, vii, 1, 2, 3, 4, 5, 10, 13, 14
thresholds, 96
time, 4, 8, 11, 14, 20, 21, 22, 23, 54, 56, 57, 77, 78, 80, 82, 83, 85, 87, 88, 89, 90, 96, 97, 99, 100, 102, 106, 107
time consuming, 100
time frame, 56, 57, 87
timing, viii, 77, 79, 80, 84, 85, 86, 105
tolerance, 12
toxicity, 13
tracking, 103, 105
trade, 21, 23, 84, 102
trade-off, 21, 23
trans, 14
transactions, 5
transformation, 22
transistor, 9
transistors, 9, 13
transition, 97, 104, 106
transmission, 4
transmits, viii, 10, 54
transparent, 12
transport, 9
transportation, 2
travel, 6
TSAT, vii, 19, 20, 21, 22, 23, 41, 57
tubular, 9

U

U.S. military, 54, 81
ultraviolet, 12
ultraviolet light, 12
uncertainty, 22
unit cost, 82, 96
United States, v, vii, 1, 2, 3, 4, 5, 6, 7, 10, 13, 14, 15, 16, 19, 53, 77, 92, 97, 98, 101

V

vacuum, 83, 84
validity, 86
van der Waals, 13
van der Waals forces, 13
vector, 9
vehicles, viii, 10, 97, 100, 102, 105, 106, 108
vibration, 10, 105
voice, 102
volatility, 84, 87
vulnerability, 2, 3, 5, 6, 7, 13

W

war, 2
warfare, 3
weapons, vii, 1, 3, 4, 5, 6, 7, 10, 13, 14, 19, 22
wear, 10
weather satellites, 6
wires, 9
workforce, 22
working groups, 21
writing, 8

Y

yield, 2